# SpringerBriefs in Molecular Science

## Green Chemistry for Sustainability

*Series Editor*

Sanjay K. Sharma

For further volumes:
http://www.springer.com/series/10045

# SpringerBriefs in Molecular Science

## Green Chemistry for Sustainability

Xuebin Yin · Linxi Yuan
Editors

# Phytoremediation and Biofortification

Two Sides of One Coin

 Springer

*Editors*
Xuebin Yin
Se Lab at University of Science
  and Technology of China
Suzhou, Jiangsu
China

Linxi Yuan
Advanced Laboratory for Selenium
  and Human Health
Suzhou Institute for Advanced Study
  University of Science and Technology
  of China
Suzhou, Jiangsu
China

ISSN 2191-5407          ISSN 2191-5415   (electronic)
ISBN 978-94-007-1438-0   ISBN 978-94-007-1439-7   (eBook)
DOI 10.1007/978-94-007-1439-7
Springer Dordrecht Heidelberg New York London

Library of Congress Control Number: 2012942919

Printed on acid-free paper

Springer is part of Springer Science+Business Media (www.springer.com)

# Preface

It's well-know natural mineral elements could show both toxicity and nutrient benefits, depending on their doses. In the past decades, phytoremediation was introduced to remove excessive mineral pollutants from soil with green plants and biofortification was another innovative biotechnology raising the mineral level in human foods. Because of different aims, the researchers get used to separately develop and utilize these two bio-technologies in their fields. Actually, they are two sides of one coin and could be closely integrated, especially for essential mineral trace elements, such as Fe, I, Cu, Zn, and Se. In this book, authors reviewed two pathways to connect phytoremediation and biofortification, as previously proposed by several international groups. First, the plant materials from phytoremediation can be further used as supplementary sources of mineral nutrients. Second, the micronutrient-laden plant materials can be made as green fertilizers to increase concentrations of micronutrient in agricultural soils. In 2009, I led a Chinese research group to make the roadmap of agricultural technology to 2050 in China, which raise a new conception on functional agriculture to produce nutraceutical foods. This trend has encouraged more studies to focus on integrating advanced biofortification and phytoremediation technologies in the practice. I believe this novel insight would determinately benefit the works in environmental remediation and micronutrient fields.

Prof. Qiguo Zhao

Chinese Academy of Sciences

# Acknowledgments

This work was funded by National Science Foundation of China (NSFC), Chinese Academy of Science (CAS), University of Science and Technology of China (USTC), and USTC-Suzhou Institute for Advanced Study. The editors and contributors wish to thank the attendants in 1st/2nd International Conference on Selenium in the Environment and Human Health (2009, 2011) for their constructive discussions and suggestions.

# Contents

# Contributors

**De Bi** Advanced Laboratory for Selenium and Human Health, Suzhou Institute for Advanced Study, University of Science and Technology of China, Suzhou 215123, Jiangsu, China

**Yang Huang** Advanced Laboratory for Selenium and Human Health, Suzhou Institute for Advanced Study, University of Science and Technology of China, Suzhou 215123, Jiangsu, China

**Tianyu Lei** Advanced Laboratory for Selenium and Human Health, Suzhou Institute for Advanced Study, University of Science and Technology of China, Suzhou 215123, Jiangsu, China

**Fei Li** Advanced Laboratory for Selenium and Human Health, Suzhou Institute for Advanced Study, University of Science and Technology of China, Suzhou 215123, Jiangsu, China

**Zhiqing Lin** Environmental Sciences Program and Department of Biological Sciences, Southern Illinois University, Edwardsville, IL 62026, USA

**Ying Liu** Advanced Laboratory for Selenium and Human Health, Suzhou Institute for Advanced Study, University of Science and Technology of China, Suzhou 215123, Jiangsu, China

**Zhangmin Wang** Advanced Laboratory for Selenium and Human Health, Suzhou Institute for Advanced Study, University of Science and Technology of China, Suzhou 215123, Jiangsu, China

**Xuebin Yin** Advanced Laboratory for Selenium and Human Health, Suzhou Institute for Advanced Study, University of Science and Technology of China, Suzhou 215123, Jiangsu, China; School of Earth and Space Sciences, University of Science and Technology of China (USTC), Hefei 230026, Anhui, China, e-mail: xbyin@ustc.edu.cn

**Linxi Yuan** Advanced Laboratory for Selenium and Human Health, Suzhou Institute for Advanced Study, University of Science and Technology of China, Suzhou 215123, Jiangsu, China

**Li Zhao** Advanced Laboratory for Selenium and Human Health, Suzhou Institute for Advanced Study, University of Science and Technology of China, Suzhou 215123, Jiangsu, China

**Yuanyuan Zhu** Advanced Laboratory for Selenium and Human Health, Suzhou Institute for Advanced Study, University of Science and Technology of China, Suzhou 215123, Jiangsu, China

# Chapter 1
# Phytoremediation and Biofortification: Two Sides of One Coin

## Xuebin Yin, Linxi Yuan, Ying Liu and Zhiqing Lin

**Abstract** Phytoremediation is a biotechnology to clean the contaminated sites by toxic elements (e.g. Cd, Cu, Zn, As, Se, Fe) via plant breeding, plant extracting, and plant volatilizing. Biofortification is an agricultural process that increases the uptake and accumulation of trace mineral nutrients (Fe, I, Cu, Zn, Mn, Co, Cr, Se, Mo, F, Sn, Si, and V) in staple crops through plant breeding, genetic engineering, or manipulation of agricultural practices. However, these two biotechnologies could be connected closely just like two sides of one coin. Actually, plant materials produced from phytoremediation could be used as supplementary sources for foods, animal feedstuff for fortified meat, or green fertilizers for fortified agricultural products. Furthermore, the transgenic technology will substantially increase their accumulation of micronutrient elements in plants or staple crops, which could be used for phytoremediation and biofortification, respectively. Future work will be needed to phytoremediate and biofortify multiple micronutrients, and then integrate both.

**Keywords** Phytoremediation · Biofortification · Integration · Micronutrition

Phytoremediation is the use of plants and their associated microbes for environmental cleanup, which has gained public acceptance in the past 15 years as a cost-

X. Yin (✉)
School of Earth and Space Science, University of Science and Technology of China (USTC), Hefei 230026, Anhui, China
e-mail: xbyin@ustc.edu.cn

X. Yin · L. Yuan · Y. Liu
Advanced Laboratory for Selenium and Human Health, Suzhou Institute for Advanced Study, University of Science and Technology of China,
Suzhou 215123, Jiangsu, China

Z. Lin
Environmental Sciences Program and Department of Biological Sciences,
Southern Illinois University, Edwardsville, IL 62026, USA

X. Yin and L. Yuan (eds.), *Phytoremediation and Biofortification*,
SpringerBriefs in Green Chemistry for Sustainability,
DOI: 10.1007/978-94-007-1439-7_1, © The Author(s) 2012

competitive and nondestructive green technology. Phytoremediation provides an alternative for engineering-based physical and chemical remediation methods. Meanwhile, biofortification is an agricultural process that increases the uptake and accumulation of mineral nutrients in agricultural products through plant breeding, genetic engineering, or manipulation of agricultural practices. The development and uses of biofortified agricultural products have recently become a promising strategy to increase the dietary nutrient intake for humans. Indeed, both phytoremediation and biofortification technologies are based on the phytoextraction process that involves plant uptake, accumulation, and transformation of nutrient elements from soil (Zhao and McGrath 2009). Although phytoremediation and biofortification have different goals, these two processes sometimes can be closely connected. This chapter discusses the applicability of different mineral nutrients (e.g., selenium, iron, and zinc) and toxic metals (such as cadmium and copper) in several suitable plant species. Both phytoextraction and biofortification have focused on enhancing the efficiency of elemental uptake and accumulation in plants. There is a strong need for better understanding the processes that affect element bioavailability, rhizosphere processes, plant uptake, translocation, distribution, and transformation in soil–plant systems. All these processes are essentially important for successful implementation of phytoremediation and biofortification strategies. In the future, phytoremediation of contaminated agricultural water and soil and biofortification of nutritionally important trace elements shall be integrated to meet the different goals of the phytotechnologies. Indeed, in some cases, phytoremediation and biofortification processes are the two sides of one coin. In this chapter, we will address this emerging concept and discuss some of the environmental and human health concerns associated with the processes of phytoremediation and biofortification.

## 1.1 Essential Micronutrient Elements for Humans

There are 20 mineral elements that are essential for human health (Vander et al. 2001), including 7 major mineral elements (Ca, P, K, S, Na, Cl, and Mg) and 13 trace elements (Fe, I, Cu, Zn, Mn, Co, Cr, Se, Mo, F, Sn, Si, and V). These elements cannot be synthesized by the body and must be continuously supplied from foods. The main physiological functions and recommended nutritional intake (RNI) and upper limit (UL) of the essential mineral trace elements are shown in Table 1.1. Because concentrations of these essential elements in soil vary substantially, plant-derived foods contain different contents of those 13 essential trace elements. When foods are lacking in one of the essential mineral nutrients in a region, local residents will suffer from malnutrition which will result in health problems. The international micronutrient organization reported that malnutrition or so-called "hidden hungry" affected one in three people worldwide. For example, approximately 2/3 Chinese dietary selenium intake is about 40 microgram per day, which is significantly lower than the recommended selenium intake value of 55 μg per day according to the World Health Organization (WHO); about one half of the Chinese population has dietary iron

**Table 1.1** Main functions, RNI, and UL of 13 mineral trace elements essential for humans

| Elements | Main functions | RNI [a] (mg/day) | UL [a] (mg/day) |
|---|---|---|---|
| Iron | a. Important part of hemoglobin<br>b. Participates in the nitrogen body exchange and breathing process<br>c. Catalyzes $\beta$, carotene into vitamin A<br>d. Induces antibodies synthesis and enhances immunity | 15 | 50 |
| Iodine | a. An essential constituent of the thyroid hormones thyroxine<br>b. Promoting growth and development of humans | 0.15 | 1.0 |
| Zinc | a. Participates in the synthesis and degradation of carbohydrates, lipids, proteins, and nucleic acids.<br>b. Promotes children's intellectual development<br>c. Accelerates teenagers' growth<br>d. Affects the palate and appetite<br>e. Affects male fertility | 15 | 45 |
| Selenium | a. Enhances immunity<br>b. Anti-aging<br>c. Inhibit cancer<br>d. Protects the heart<br>e. Antagonist heavy metal | 0.05 | 0.4 |
| Copper | a. An important component of proteins and enzymes<br>b. Closely related to human body hematopoiesis<br>c. Affects antioxidant ability of body | 2.0 | 8.0 |
| Molybdenum | a. An important component of xanthine oxidase and aldehydes oxidase<br>b. Takes part in the electronic transmission of cell<br>c. Restrains the breeding of virus in cell | 60 | 280 |
| Chromium | a. Promotes protein metabolism and body growth<br>b. Influences lipid metabolism<br>c. An important part of glucose tolerance factor | 0.05 | 0.5 |
| Silicon | a. Plays an essential role in the development of bone<br>b. Participates in the metabolism of the polysaccharide | – | – |
| Nickel | a. Is a component of hydrogenated enzyme<br>b. Promotes the formation of insulin<br>c. Lowers blood glucose | 0.1 | 0.6 |
| Cobalt | a. Is a component of Vitamin B12<br>b. Participates in hemoglobin synthesis | – | – |
| Vanadium | a. Maintains normal metabolism of fat<br>b. Is a constituent of nucleic acid<br>c. Promotes the growth of bones and teeth | 0.03 | 10 |
| Fluoride | a. Plays an important part in the growth of bones and teeth | 1.5 | 3.0 |
| Tin | a. Has a function in the tertiary structure of proteins or other biosubstances<br>b. Is used as catalyst for polymerization, transesterification, and olefin condensation reactions | 15 | – |

*Note* [a] Means male adult (ages: 18–50)

intake of about 3 microgram per day, which is also lower than the recommended value of 15 microgram by WHO. Globally, there are 80 % children with zinc deficiency for their rapid growth. To increase dietary intake of essential micronutrient elements, biofortification was proposed and regarded as an economic and promising approach for developing countries. Scientists worldwide have published many research papers in the past two decades (http://apps.webofknowledge.com). These studies help with better understanding on the biofortification technology, potential health effects, and food safety regulations.

While the 13 trace elements are considered essential for human health, they can also become environmental pollutants due to their excessive levels in soils. To protect the environment and to minimize local environmental risk, the polluted sites need to be remediated. In the past decades, phytoremediation was introduced as a successful green biotechnology (Terry and Bañuelos 2000). However, one of the difficulties that we are facing in phytoremediation management is to deal with the large volume of polluted plant waste materials harvested from phytoremediation sites. Different management options have been discussed by researchers regarding the disposal of plant waste materials, including landfill and incineration. But, none of them are sustainable or environmental-friendly. Generally, the phytoremediation plant waste materials contained high concentrations of the pollutant trace elements, and were potentially toxic to humans and wildlifes via direct consumption exposure. One may hypothesize that those polluted plant materials can be used to produce agricultural crops that are enriched with the essential micronutrients. For example, selenium, zinc, and iron-laden plant materials can be used as valuable sources of nutrients in agricultural production systems. However, the plant materials should not contain high levels of other toxic heavy metals, such as cadmium, arsenic, and mercury. All these will be further discussed in this book.

## 1.2 Can Phytoremediation Plants Become Sources of Human Micronutrient Elements?

Some plants are able to concentrate large amounts of specific trace elements in their leaves or stems (Robinson Brett et al. 2009; Schwitzguebel et al. 2009). This natural process has been applied to remediate the metal-polluted soil and water. As a result, the plant materials produced from phytoremediation can be further used as supplementary sources of mineral nutrients to produce food or feedstuff or functional biofortified agricultural products. Micronutrient-laden plant materials can be used as green fertilizers to increase concentrations and bioavailability of micronutrient trace elements in agricultural soils, or used as animal feed to increase dietary intake of micronutrients by animals which further enhances the accumulation of micronutrients in animal originated food products. Therefore, to integrate phytoremediation and biofortification processes, the chemical composition of phytoremediation plant materials is of utmost concern. The contents of toxic metals accumulated in plant materials will essentially jeopardize the use of phytoremediation plant materials for biofortification. It is critically important to screen and select the right plant species and acceptable phytoremediation field sites to implement the integration of phytoremediation and biofortification strategies. In general, there are two very basic requirements to meet this goal: first, the selected plants should be edible; second, the edible part of the plant should accumulate more micronutrients, but very less toxic trace elements. For example, Indian mustard was used for phytoremediation of selenium-contaminated water and soil in agricultural lands of the San Joaquin Valley, Central California. The selenium-laden mustard plant materials have also been used as biofortified selenium supplement for animals and humans (Bañuelos et al. 2007, 2009, 2011; Turan and Bringue 2007; Hamlin and Barker 2008). Additionally, researchers have also applied genetic engineering technology to substantially increase plant accumulation of micronutrient elements (Bañuelos and Lin 2009; Manohar et al. 2011).

Phytoremediation commonly selects the plant species that accumulate more pollutants in shoots, and focuses on the phytoextraction or remediation efficiency, while biofortification focuses on increasing micronutrient contents in crops. If the biofortified materials are directly utilized as food supplements to increase human mineral dietary intake, the phytoremediation plant should be edible.

## 1.3 Managing Toxic Metals in Plant Tissues

When phytoremediation plant materials can be used as sources of nutrient trace elements for biofortification, the connection between phytoremediation and biofortification could be still problematic. Since the contaminated sites are often contaminated with multi-pollutants, including toxic cadmium, mercury, and arsenic, the use of phytoremediation plant materials for biofortification becomes more difficult.

Previous studies indicated that the manipulation of soil physicochemical properties, including soil pH, total organic carbon (TOC), fulvic acid (such as citrate), and chelate can change the uptake and accumulation of various nutrient elements by plants. Some organic acids produced in the rhizosphere may play important roles in determining bioavailability of mineral trace elements in the soil and affect nutrient uptake efficiency via roots.

## 1.4 Future Research Needs

Recent studies on phytoextraction have been partially focused on the development of biofortified agricultural products. New efforts have been made to integrate the phytoremediation with biofortification processes. There are still many scientific questions that have not been answered. The future research shall investigate the feasibility of biofortification of multiple micronutrients, such as increasing accumulation of both selenium and zinc in crops or vegetables (Srivastava et al. 2009; Zhu et al. 2009) For example, *Thlaspi caerulescens* is zinc hyperaccumulator and *Stanleya pinnata* is selenium hyperaccumulator. The application of these two plant materials as green manures in agricultural soils could significantly increase the total content and bioavailability of both zinc and selenium in the soil, and therefore, enhance the accumulation of zinc and selenium in the edible portion of crops.

# References

Bañuelos GS, LeDuc Danika L, Pilon-Smits Elizabeth AH et al (2007) Transgenic Indian mustard overexpressing selenocysteine lyase or selenocysteine methyltransferase exhibit enhanced potential for selenium phytoremediation under field conditions. Environ Sci Technol 41(2):599–605

Bañuelos GS, Lin Z-Q (eds) (2009) Development and uses of biofortified agricultural products. CRC Press, Boca Raton, p 297

Bañuelos GS, Lin Z-Q, Yin XB (eds) (2009) Selenium deficiency, toxicity, biofortification and human health. USTC press, Hefei, p 10

Bañuelos GS, Lin Z-Q, Yin XB, Duan N (eds) (2011) Selenium: global perspectives of impacts on humans, animals and the environment. USTC press, Hefei, p 10

Hamlin RL, Barker AV (2008) Nutritional alleviation of zinc-induced iron deficiency in Indian mustard and the effects on zinc phytoremediation. J Plant Nutr 31(12):2196–2213

Manohar M, Shigaki T, Hirschi KD et al (2011) Plant cation/H(+) exchangers (CAXs): biological functions and genetic manipulations. Plant Biol 13(4):561–569

Vander A, Sherman J, Luciano D (2001) Human physiology: the mechanisms of body function, 8th edn. McGraw-Hill Higher Education, New York

Robinson BH, Banuelos G, Conesa HM et al (2009) The phytomanagement of trace elements in soil. Crit Rev Plant Sci 28(4):240–266

Schwitzguebel J-P, Kumpiene J, Comino E et al (2009) From green to clean: a promising and sustainable approach towards environmental remediation and human health for the 21(st) century. Agrochimica 53(4):209–237

Srivastava S, Mishra S, Dwivedi S et al (2009) Evaluation of zinc accumulation potential of Hydrilla verticillata. Biol Plant 53(4):789–792

Terry N, Bañuelos GS (eds) (2000) Phytoremediation of trace elements in contaminated water and soil. Lewis Publishers, Boca Raton, p 389

Turan M, Bringue A (2007) Phytoremediation based on canola (*Brassica napus L.*) and Indian mustard (*Brassica juncea L.*) planted on spiked soil by aliquot amount of Cd, Cu, Pb, and Zn. Plant Soil Environ 53(1):7–15

Zhao F-J, McGrath SP (2009) Biofortification and phytoremediation. Curr Opin Plant Biol 12(3):373–380

Zhu YG, Pilon-Smits EAH, Zhao FJ et al (2009) Selenium in higher plants: understanding mechanisms for biofortification and phytoremediation. Trends Plant Sci 14(8):436–442

# Chapter 2
# Selenium in Plants and Soils, and Selenosis in Enshi, China: Implications for Selenium Biofortification

Linxi Yuan, Xuebin Yin, Yuanyuan Zhu, Fei Li, Yang Huang, Ying Liu and Zhiqing Lin

**Abstract** The total selenium (Se) content of soils in Enshi, China, the so-called "World Capital of Selenium", is concentrated in a range of 20–60 mg/kg DW which is approximately 150–500 times greater than the average Se content (0.125 mg/kg DW) in Se-deficient areas and approximately 50–150 times greater than that (0.40 mg/kg DW) in Se-enriches areas in China, respectively. However, the distribution of Se in soils is greatly uneven with some exceptionally high contents of more than 100 mg/kg DW, which is very likely caused by the micro-topographical features and leaching conditions. Among the 14 plant species in Enshi, *Adenocaulon himalaicum* has the highest contents of Se from 299 to 2,278 (mean 760) mg/kg DW in the leaf, from 268 to 1,612 (mean 580) mg/kg DW in the stem, from 227 to 8,391 (mean 1,744) mg/kg DW in the root, and therefore was identified as a secondary Se-accumulating plant. Furthermore, the SeCys2 fraction was predominant in the tissues with a proportion of 70–98 %, which is quite different from other Se-accumulating plants, e.g., garlic, onion, and broccoli. Although the Se concentration in resident foods and the daily Se intake decreased significantly from 1963 to 2010 in Enshi, the present daily Se intake (575 μg/d) is still above the recommended maximum safe intake of 550 μg/d, which indicates that there may be potential risk for selenosis in Enshi. Both Se distributions in soils

L. Yuan · X. Yin (✉) · Y. Zhu · F. Li · Y. Huang · Y. Liu
Advanced Laboratory for Selenium and Human Health, Suzhou Institute
for Advanced Study, University of Science and Technology of China,
Suzhou, 215123, Jiangsu, China
e-mail: xbyin@ustc.edu.cn

X. Yin
School of Earth and Space Sciences, University of Science and Technology of China
(USTC), Hefei, Anhui 230026, China

Z. Lin
Environmental Sciences Program and Department of Biological Sciences, Southern Illinois
University, Edwardsville, IL 62026, USA

X. Yin and L. Yuan (eds.), *Phytoremediation and Biofortification*,
SpringerBriefs in Green Chemistry for Sustainability,
DOI: 10.1007/978-94-007-1439-7_2, © The Author(s) 2012

and plants and human daily Se intakes obviously indicate that Enshi, China should be Se-phytoremediated to decrease the risk for selenosis there. Fortunately, Se-biofortification was taken as an effective method to overcome this problem. Hopefully, Enshi, China is moving on a natural field-scale trial for integration of Se-phytoremediation and Se-biofortification.

**Keywords** Selenium · Enshi, China · Soil · Plant · Selenium intake

## 2.1 Introduction

Selenium (Se), discovered by the Swedish chemist Jakob Berzelius in 1817, is a metalloid and states in group VIA with an atomic weight of 78.96. Selenium has five valence states in nature, including selenide (2−), elemental Se (0), thioselenate (2+), selenite (4+), and selenate (6+). Selenium is an essential nutrient for humans and animals to form important selenoproteins, including glutathione peroxide, thioredoxin reductase (Terry et al. 2000; Zhu et al. 2009). In 1973, Se was found to be involved in forming the active center of glutathione peroxidase and thioredoxin reductase enzymes; these enzymes play important roles in reducing certain oxidized molecules in animals (Liu et al. 2010).

The range between the beneficial and harmful concentrations of Se is quite narrow; the minimal Se nutrition levels for animals is about 0.05–0.10 mg/kg dry forage feed, while the toxic exposure level is 2–5 mg/kg dry forage (Wilber 1980; Wu et al. 1996). The World Health Organization (WHO) recommended the required dietary intake of Se to be 50–200 µg/day for adults (WHO 1987). Two well-known endemic diseases, Keshan Disease (a degenerative heart disease bursting out in Kesha, Heilongjiang, China) and Kaschin-Beck Disease (an osteoarthropathy which causes deformity of the affected joints) were linked to soil Se-deficiency and low Se daily intake (Tan and Huang 1991; Tan et al. 2002). However, because of long-term exposure to high levels of Se, Se toxicological symptoms, including hair and nail loss and nervous system disorders, extensively occurred in inhabitants in two notable Se-enriched areas, Enshi, Hubei, China and Ziyang, Shanxi, China (Yang et al. 1981a, b, 1983; Mei 1985; Li et al. 2011).

## 2.2 Enshi, the World Capital of Selenium

Enshi (E 108°23′12″–110°38′08″, N 29°07′10″–31°24′13″) is a national minority autonomous prefecture located in Northwestern Hubei Province, China (Fig. 2.1). From early 1930 to 1960, people living in Yutangba, Huabei, and Shadi villages of Enshi, experienced loss of hair and nails, showing the typical symptoms of Se toxicity (Zhang et al. 1998). For instance, 19 of 23 local inhabitants in Yutangba showed visible Se poisoning symptoms and all livestock in the village died in 1963. Subsequently, villagers were evacuated from their homes in Yutangba (Mao et al. 1990, 1997). After the occurrence of the incident, selenosis has become a matter of concern for local governments, scientists, and Se endemic disease investigators (Tan and Huang 1991; Zhang et al. 1998; Wang and Gao 2001; Tan et al. 2002). Among the studies carried out in Enshi, Yu (1993) reported the discovery of Se mines in Yutangba village, with a high Se content of 8,500 mg/kg. The Se mines were formed in Maokou, in the late Permian period with a thickness of 13 m, which were the "culprits" for the selenosis observed in the Yutangba village (Fig. 2.1).

## 2.3 Selenium in Plants

To investigate the characteristics of Se pollution in the seleniferous areas in Enshi, the dominant plants and their underlying soils were collected in Yutangba, including 14 species and 8 classes (Table 2.1).

The plant samples were rinsed in deionized water, and most of the plants were separated into root, stem, and leaf for Se content analysis, except for *Adenocaulon himalaicum, Elsholtzia splendens, Trifolium repens, Lycodium clavaturn, Polygonum hydropiper,* and *Rumex japonicas* that do not have true stems and were separated only into roots and shoots. The Se content in *Mosla dianthera* seed was also determined.

Our results showed that *Adenocaulon himalaicum* had exceptional high concentrations of Se with 563.60 mg/kg DW in the root and 1,317.46 mg/kg DW in the leaf, followed by *Medicago sativa* that accumulated Se concentrations of 150.96, 154.40, and 168.14 mg/kg DW in root, stem, and leaf, respectively. Furthermore, the leaves of *Sedum sarmentosum, Trifolium repens,* and *Mosla dianthera* had relatively higher concentrations of Se, compared with the Se contents in roots. The Se translocation factor (i.e., the ratios of shoot to root Se concentrations) of 4.5 was the highest in *Trifolium repens*, while the Se concentration in the root was only 17.65 mg/kg DW. For *Mosla dianthera* and *Sedum sarmentosum,* the Se translocation factor was greater than 2.

## 2.4 Selenium in Soils

The sequential chemical-extraction technique is a conventional method to evaluate the geochemical behavior of trace elements in soil (Sharmasarkar and Vance 1995; Mao and Xing 1999; Zhang et al. 2002). Se has several chemical forms in soil, such as $Se^0$, $SeO_3^{2-}$, $SeO_4^{2-}$, and organic Se. In the sequential chemical extraction, the fraction by water-extraction is called the water-soluble Se (Fraction 1), the fraction by $KH_2PO_4$–$K_2HPO_4$-extraction is called the exchangeable Se (Fraction 2), the fraction by HCl-extraction is called the acid-soluble Se (Fraction 3), the fraction

**Fig. 2.1** Sketch map showing the location of Enshi, China (**a**) and the sketch geological map of Yutangba, Enshi, and the Se ore vents were marked (**b**)

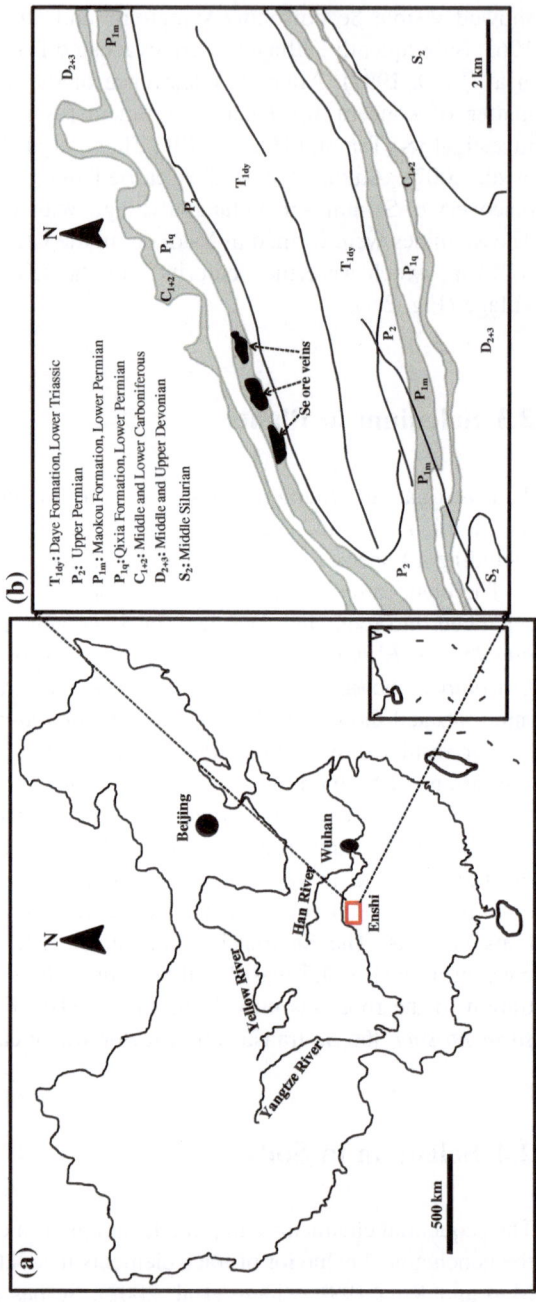

by $K_2S_2O_8$-extraction is called the organic-bound Se (Fraction 4), and the remaining fraction is called the residue Se (Fraction 5). Among those fractions, the water-soluble Se and the exchangeable Se are considered as bioavailable Se, the

**Table 2.1** The concentrations of Se in roots, stems, and leaves of plants from Enshi (mg/kg DW) and the ratio between the calculated shoot (Stem + Leaf) and the root

| No. | Latin name | Class | Root | Stem | Leaf | Shoot/Root |
|---|---|---|---|---|---|---|
| 1 | *Adenocaulon himalaicum* | Asteraceae | 563.60 | / | 1,317.46 | 2.34 |
| 2 | *Siegesbeckia orientalis* | Compositae | 112.29 | 84.61 | 4.92 | 0.80 |
| 3 | *Erigeron annuus* | Compositae | 18.03 | 14.83 | 13.56 | 1.57 |
| 4 | *Artemisia lavandulaefolia* | Compositae | 1.62 | 0.39 | 1.10 | 0.92 |
| 5 | *Sedum sarmentosum* | Crassulaceae | 43.12 | 19.94 | 99.22 | 2.76 |
| 6 | *Miscanthus sinensis* | Gramineae | 24.64 | 14.79 | 25.44 | 1.63 |
| 7 | *Miscanthus purpurascens* | Gramineae | 123.88 | 25.56 | 62.64 | 0.71 |
| 8 | *Mosla dianthera* | Labiatae | 8.91 | 7.24 | 12.84(Seed) | 2.25 |
| 9 | *Elsholtzia splendens* | Labiatae | 19.86 | / | 11.83 | 0.60 |
| 10 | *Trifolium repens* | Leguminosae | 17.65 | / | 79.36 | 4.50 |
| 11 | *Medicago sativa* | Leguminosae | 150.96 | 154.40 | 168.14 | 2.14 |
| 12 | *Lycodium clavaturn* | Lycopodiaceae | 1.48 | / | 1.73 | 1.17 |
| 13 | *Polygonum hydropiper* | Polygonaceae | 21.33 | / | 27.57 | 1.29 |
| 14 | *Rumex japonicus* | Polygonaceae | 18.55 | / | 31.66 | 1.71 |

HCl-soluble Se and the organic-bound Se are regarded as the transferable Se, and the residue Se are regarded as the un-bioavailable Se. Overall, the bioavailable Se content in soil is the key factor for the Se accumulation in plant, and the transferable Se content in soil provides a potential Se source for plant uptake (Zhao et al. 2005; Zhu et al. 2008a).

The results of the total Se and the fractions of sequential chemical-extraction on Se in soils are shown in Table 2.2. The total Se contents varied from 3 to 4 mg/kg DW in the underlying soils of *Lycodium clavaturn* and *Artemisia lavandulaefolia* to 100–436 mg/kg DW in the underlying soil of *Miscanthus sinensis, Sedum sarmentosum,* and *Miscanthus purpurascens.* But the total Se concentrations in most of the soil samples collected in Yutangba contained 20–60 mg/kg DW, which is approximately 150–500 times greater than the average soil Se content (about 0.125 mg/kg DW) in Se-deficient areas (Tan and Huang 1991; Tan et al. 2002). When compared with the total soil Se concentrations in other Se-enriched areas worldwide, such as 3 mg/kg DW in China and 2.41 mg/kg DW in the western U.S., the soil total Se concentrations in Yutangba of Enshi was 10–30 times higher (Presser et al. 1994). It should be pointed out that the soil samples containing very high Se concentrations, such as the underlying soils of *Sedum sarmentosu* and *Miscanthus purpurascens,* were collected from the discarded Se-coal spoils. Although the local lithological differences could result in considerable variation in soil Se distribution (Fordyce et al. 2000), it is likely that micro-topographical features and hydrological conditions were the primary factors affecting the soil Se content and distribution in the study area (Zhu and Zheng 2001).

The fractionation analysis of Se in the vegetated soils revealed that the total Se concentration in the fraction 1 ranged from 1 to 2 mg/kg DW with lower concentrations in the underlying soils of *Lycodium clavaturn* (0.30 mg/kg DW) and *Adenocaulon himalaicum* (0.45 mg/kg DW), and with a higher concentration

**Table 2.2** Fractional partitioning of Se in the underlying soils (mg/kg DW)

| No. | Fraction 1 | Fraction 2 | Fraction 3 | Fraction 4 | Fraction 5 | Total Se |
|-----|-----------|-----------|-----------|-----------|-----------|----------|
| 1 | 0.45 | 1.81 | 8.44 | 9.13 | / | 19.82 |
| 2 | 1.28 | 1.58 | 6.31 | 6.46 | 29.70 | 45.33 |
| 3 | 0.91 | 1.20 | 7.03 | 8.87 | 24.02 | 42.03 |
| 4 | 0.64 | 0.82 | 0.92 | 1.17 | 0.21 | 3.76 |
| 5 | 1.04 | 3.51 | 12.70 | 10.41 | 111.24 | 138.90 |
| 6 | 2.30 | 1.77 | 9.86 | 14.49 | 72.38 | 100.80 |
| 7 | 6.85 | 15.41 | 82.52 | 62.22 | 268.70 | 435.70 |
| 8 | 1.35 | 1.42 | 2.91 | 4.59 | 17.84 | 28.11 |
| 9 | 0.98 | 2.79 | 4.12 | 5.16 | 5.09 | 18.14 |
| 10 | 1.09 | 0.91 | 2.28 | 4.17 | 37.17 | 45.62 |
| 11 | 1.81 | 1.84 | 3.23 | 3.79 | 32.95 | 43.62 |
| 12 | 0.30 | 0.66 | 0.77 | 0.99 | 0.46 | 3.18 |
| 13 | 0.87 | 2.32 | 12.13 | 8.32 | 37.07 | 60.71 |
| 14 | 1.19 | 2.18 | 13.30 | 10.94 | 51.47 | 79.08 |

in the underlying soil of *Miscanthus purpurascens* (6.85 mg/kg DW). The Se in the fraction 2 was in a range of 1–3 mg/kg DW, with a low value in the underlying soil of *Lycodium clavaturn* (0.66 mg/kg DW) and a high value in the underlying soil of *Miscanthus purpurascens* (15.41 mg/kg DW). The Se distribution in fractions 3 (1–4 mg/kg DW) and 4 (1–5 mg/kg DW) were different compared with the Se distribution in other fractions. Relatively low concentrations of Se in fractions 3 and 4 were found in the underlying soils of *Artemisia lavandulaefolia, Mosla dianthera, Elsholtzia splendens, Trifolium repens, Medicago sativa,* and *Lycodium clavaturn.* In contrast, higher concentrations of Se in fractions 3 (7–13 mg/kg DW) and 4 (6.5–14.5 mg/kg DW) were observed in the underlying soils of *Adenocaulon himalaicum, Siegesbeckia orientalis, Erigeron annuus, Sedum sarmentosum, Miscanthus sinensis, Polygonum hydropiper,* and *Rumex japonicas.* Very high Se concentrations of 382.52 mg/kg DW in fraction 3 and of 62.22 mg/kg DW in fraction 4 were determined in the underlying soil of *Miscanthus purpurascens.* For fraction 5, Se concentrations were very low in the underlying soils of *Artemisia lavandulaefolia* (0.21 mg/kg DW), *Lycodium clavaturn* (0.46 mg/kg DW), and *Elsholtzia splendens* (5.09 mg/kg DW). Concentrations of Se in fraction 5 ranged from 20 to 70 mg/kg DW, with an exception in the underlying soils of *Sedum sarmentosum* (111.24 mg/kg DW) and *Miscanthus purpurascens* (268.70 mg/kg DW).

The percentages of Se distribution among different fractions in the underlying soils are shown in Fig. 2.2. Overall, Se in fraction 5 accounted for 40–80 % of the total Se, 10–20 % in fractions 3 and 4, and less than 3 % in fractions 1 and 2. Therefore, the proportion of bioavailable Se was <5 % in the underlying soils in Enshi. However, the proportion of transferable Se was relatively high (20–40 %), which could be used by plants for uptake. As for fraction 5, un-bioavailable Se was predominant in the vegetated soils.

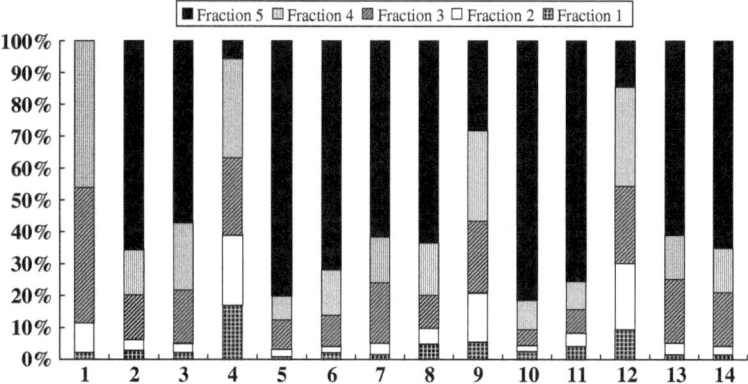

**Fig. 2.2** The percentage of the fractional partitioning of Se in the underlying soils

**Table 2.3** Variations of Se concentrations in soil and stream water in Enshi during 1963–2010

| Se in soil (mg/kg dry weight) mean (min–max) | Se in stream water (μg/L) mean (min–max) | Sampling time (year) | References |
|---|---|---|---|
| 6.83 | / | 1963 | Mao et al. (1997) |
| 9.68 (0.08–45.5) | 56 (0–158) | 1966 | Yang et al. (1981a, b) |
| 3.45 (1.92–4.98) | / | 1987 | Mao et al. (1997) |
| 5.48 (0–11.89) | / | 1989 | Zheng et al. (1993) |
| 4.06 (2.82–5.30) | / | 1992 | Zhu and Zheng (2001) |
| 4.99 (2.61–7.37) | 40.4 | 1996 | Fordyce et al. (2000) |
| 4.75 (0–12.18) | 58.4 (41.6–75.2) | 1999 | Zhu et al. (2008b) |
| 27.81 (3.76–79.08) | 52.66 (15.13–192.70) | 2010 | This study |

In comparing with other total soil Se concentrations reported previously by other researchers from the same research areas in Yutangba village, the temporal variation of soil Se concentration is shown for the time period from 1963 to 2010 in Table 2.3. The soil Se concentration in Yutangba was 6.83 mg/kg DW in 1963 (Mao et al. 1997) and 9.68 mg/kg DW, with a range of 45.42 mg/kg in 1966 (Yang et al. 1981a, b). During the time period from 1987 to 1999, the soil Se concentrations were from 3.5 to 5 mg/kg DW. However, in a recent study conducted in 2010, Yin and his colleagues reported that the soil Se concentrations in Yutangba varied from 3.76 to 79.08 mg/kg DW, with an average of 27.81 mg/kg. Based on these Se concentrations in the soil samples collected from 1963 to 2010 by different research groups, the soil Se concentrations in Yutangba were high in 1963 and 1966, low and relatively stable from 1987 to 1999, and then were higher again in 2010. Overall, the total Se content in soils in Enshi generally varied from 4 to 25 mg/kg DW which were approximately 30–200 times greater than the average Se content (0.125 mg/kg DW) in Se-deficient areas (Fordyce et al. 2000).

The Se concentrations in stream water were 40–60 µg/L, which is approximately 4–6 times greater than the drinking water maximum concentration of 10 µg/L recommended by the World Health Organization (WHO) and the US Environmental Protection Agency (Presser et al. 1994).

## 2.5 Plant Uptake of Selenium from Seleniferous Soil in Enshi

To estimate the ability of plants to take up Se from soil, the root bioconcentration factors (BCF = $[Se]_{plant\ root}/[Se]_{soil}$) were calculated for the plant species tested in the present study (Table 2.4). *Adenocaulon himalaicum, Medicago sativa,* and *Siegesbeckia orientalis* showed relatively high root BCFs of 28.44, 3.46, and 2.48, respectively. For stem tissues, *Medicago sativa* had the highest stem BCF with 3.54, 42.30, and 14.47 for S/T ($[Se]_{plant\ stem}/[Se]_{soil}$), S/B ($[Se]_{plant\ stem}/[Se]_{bioavailable\ in\ soil}$), and S/(B + Tr) ($[Se]_{plant\ stem}/[Se]_{bioavailable\ plus\ transferable\ in\ soil}$), respectively. The stem of *Siegesbeckia orientalis* also apparently accumulated Se from the underlying soil with BCFs more than 1. Although most of the ratios of S/B were more than 1 in the other plant species, the ratios of S/T and S/(B + Tr) on them were lower than 0.3 for S/T and 0.8 for S/(B + Tr). For leaf tissues, *Adenocaulon himalaicum* and *Medicago sativa* apparently accumulated Se from the underlying soil with the ratios of L/T ($[Se]_{plant\ leaf}/[Se]_{soil}$) of 66.47 and 3.85, respectively, which displayed more transportation efficiency than that in the root and the stem, especially for *Adenocaulon himalaicum*. It should be pointed out that the leaf of *Siegesbeckia orientalis* had a very low ratio of L/T. In contrast, the leaf of *Trifolium repens* could accumulate Se from the underlying soil with a ratio of L/T of 1.74, although its root did not display this feature. The other plant species had the ratios of L/T of less than 0.5, which revealed that those plants did not prefer Se. Similar trends were found in the ratios of L/B ($[Se]_{plant\ leaf}/[Se]_{bioavailable\ in\ soil}$) and L/(B + Tr) ($[Se]_{plant\ stem}/[Se]_{bioavailable\ plus\ transferable\ in\ soil}$).

Overall, *Adenocaulon himalaicum* displayed the exceptional ability to accumulate Se in its root, stem, and leaf tissues. *Medicago sativa* was also a good Se-accumulator. *Trifolium repens* accumulates Se in its leaf part, but not in other parts.

The relationships between plant selenium accumulation and the extracted fractions in vegetated soils, and Se concentrations in different plant tissues are compiled in Fig. 2.3. The results show that the Se concentration in root significantly correlated with ($R^2 = 0.81$, $P < 0.05$) the total Se content in the soil. The sum of bioavailable and transferable Se, not total Se in the underlying soil dominated the Se content in the plant stem and the plant leaf with a high positive correlation coefficient of 0.87 and 0.81, respectively, which is different from that in plant root (Fig. 2.3).

**Table 2.4** The bioconcentration factors (BCF) of root (R), stem (S), and leaf (L) compared with the total Se content (T), the bioavailable Se content (B), and the transferable Se content (Tr) of underlying soils, respectively

| No. | Latin name | Root (R) | | | Stem (S) | | | Leaf (L) | | |
|---|---|---|---|---|---|---|---|---|---|---|
| | | R/T | R/B | R/(B + Tr) | S/T | S/B | S/(B + Tr) | L/T | L/B | L/(B + Tr) |
| 1 | *Adenocaulon himalaicum* | 23.44 | 249.38 | 28.42 | / | / | / | 66.47 | 582.95 | 66.44 |
| 2 | *Siegesbeckia orientalis* | 2.48 | 39.26 | 7.18 | 1.87 | 29.58 | 5.41 | 0.11 | 1.72 | 0.31 |
| 3 | *Erigeron annuus* | 0.43 | 8.55 | 1.00 | 0.35 | 7.03 | 0.82 | 0.32 | 6.43 | 0.75 |
| 4 | *Artemisia lavandulaefolia* | 0.43 | 1.11 | 0.46 | 0.10 | 0.27 | 0.11 | 0.29 | 0.75 | 0.31 |
| 5 | *Sedum sarmentosum* | 0.31 | 9.48 | 1.56 | 0.14 | 4.38 | 0.72 | 0.71 | 21.81 | 3.59 |
| 6 | *Miscanthus sinensis* | 0.24 | 6.05 | 0.87 | 0.15 | 3.63 | 0.52 | 0.25 | 6.25 | 0.90 |
| 7 | *Miscanthus purpurascens* | 0.28 | 5.57 | 0.74 | 0.06 | 1.15 | 0.15 | 0.14 | 2.81 | 0.38 |
| 8 | *Mosla dianthera* | 0.32 | 3.22 | 0.87 | 0.26 | 2.61 | 0.70 | 0.46 | 4.64 | 1.25 |
| 9 | *Elsholtzia splendens* | 1.09 | 5.27 | 1.52 | / | / | / | 0.65 | 3.14 | 0.91 |
| 10 | *Trifolium repens* | 0.39 | 8.83 | 2.09 | / | / | / | 1.74 | 39.68 | 9.39 |
| 11 | *Medicago sativa* | 3.46 | 41.36 | 14.15 | 3.54 | 42.30 | 14.47 | 3.85 | 46.07 | 15.76 |
| 12 | *Lycodium clavatum* | 0.47 | 1.54 | 0.54 | / | / | / | 0.54 | 1.80 | 0.64 |
| 13 | *Polygonum hydropiper* | 0.35 | 6.69 | 0.90 | / | / | / | 0.45 | 8.64 | 1.17 |
| 14 | *Rumex japonicus* | 0.23 | 5.50 | 0.67 | / | / | / | 0.40 | 9.39 | 1.15 |

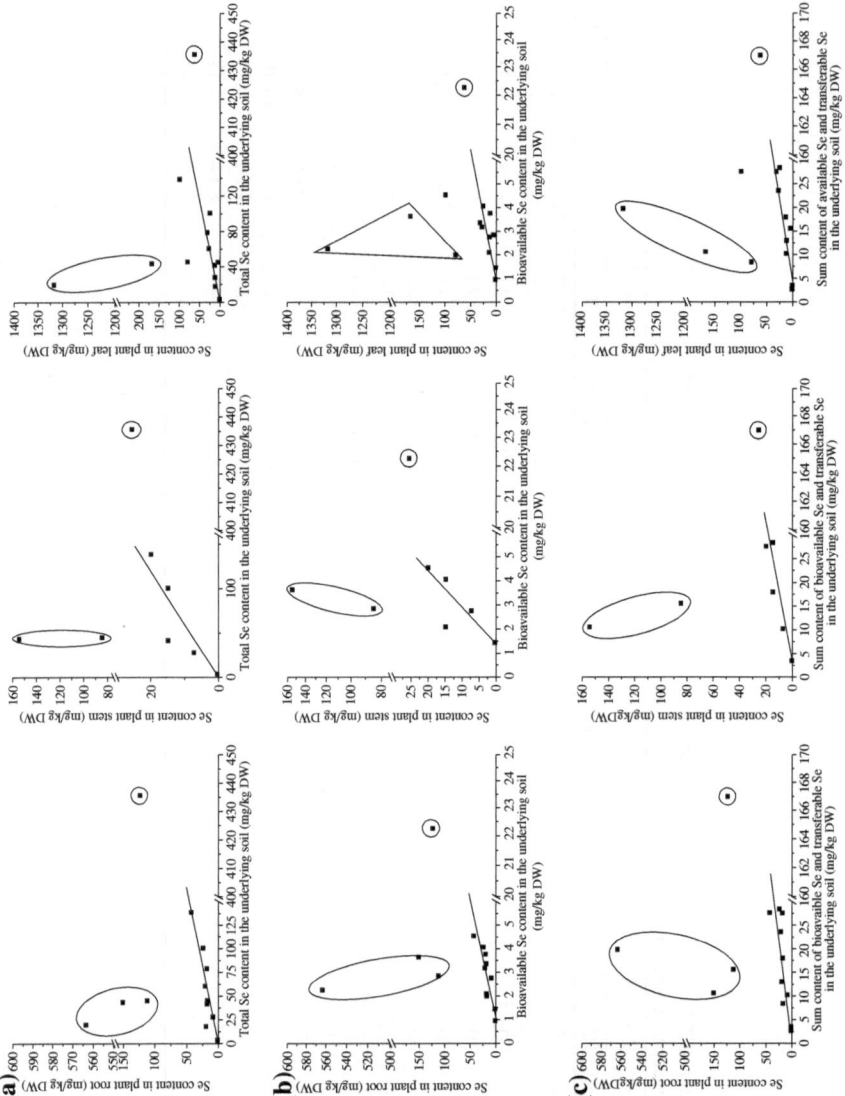

**Fig. 2.3** The relationship between the Se content of extracted fraction in the underlying soil and the Se content of plant tissue. (**a**) The total Se content of underlying soil versus the Se content of plant root, plant stem, and plant leaf, respectively; (**b**) The bioavailable Se content of underlying soil versus the Se content of plant root, plant stem, and plant leaf, respectively; (**c**) The sum content of bioavailable Se and transferable Se in underlying soil versus the Se content of plant root, plant stem, and plant leaf, respectively

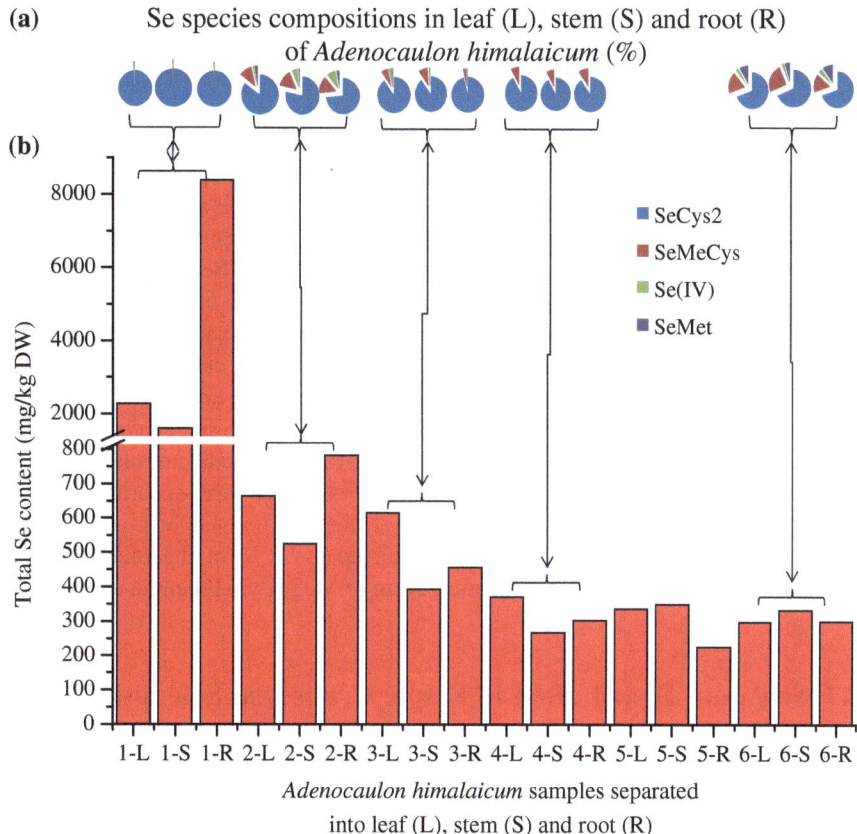

**Fig. 2.4** The Se species compositions (**a**) and the contents of total Se (**b**) in leaves, stems, and roots of *Adenocaulon himalaicum* from Enshi

## 2.6 Selenium Hyperaccumulating Plant and its Implications

Generally, Se concentrations in plants in Se-enriched soils were less than 25 mg/kg DW (Bell, Parker and Page 1992), except for a few Se-hyperaccumulator species containing over 1,000 mg/kg (Ellis and Salt 2003). Our current study shows that *Adenocaulon himalaicum* could be classified as a secondary Se-accumulating species. Figure 2.4b shows the concentrations of total Se, Selenocystine (SeCys2), Se-Methylselenocysteine (SeMeCys), and Selenomethionine (SeMet) in the leaves, stems, and roots of *Adenocaulon himalaicum*. The total Se concentrations were $760 \pm 692$ mg/kg DW in the leaf, $580 \pm 468$ mg/kg DW in the stem, and $1744 \pm 2978$ mg/kg DW in the root. Selenium speciation analysis indicated that $SeCys_2$, SeMeCys, and SeMet accounted for 70–98, 7–19, and 3–11 %, respectively, of the total Se accumulated in *Adenocaulon himalaicum* leaves (Fig. 2.4a). A similar pattern occurred in the stem and the root tissues. The proportion of SeCys$_2$ in *Adenocaulon himalaicum* increases with increasing the accumulation of total Se in the plant tissues.

In the literature, *Arabidopsis thaliana* and *B. juncea* accumulate Se mainly in the chemical form of selenate. However, when soils were supplied with selenate, garlic (*Allium sativum*), onion (*Allium cepa*), leek (*Allium ampeloprasum*), and broccoli (*Brassica oleracea*) accumulate Se primarily as SeMeCys (Beilstein et al. 1991; Kahakachchi et al. 2004; Pilon-Smits and Quinn 2010). SeMet is a common dominant Se species in most grains, such as wheat, barley, and rye (Stadlober et al. 2001). However, to the best of our knowledge, this is the first time that SeCys2 is identified as the dominant Se chemical species in higher plants. This finding will provide important insight into Se-metabolism pathways for Se hyperaccumulator species. Moreover, if this Se-hyperaccumulating plant species could be cultivated and planted widely in Enshi, it could be a good Se-supplement source for animals or humans.

Generally, nonhyperaccumulating plants use Se through S pathways. However, recent studies suggest that Se might be essential for Se hyperaccumulator species and present specialized Se-specific transporters which are separate from S movement (Feist and Parker 2001; Galeas et al. 2007). Even in a different Se hyperaccumulator species (*Stanleya pinnata*), up to 90 % of the total Se accumulated in plant tissues is in the chemical form of MeSeCys (Freeman et al. 2006).

## 2.7 Selenium Distribution in Staple Crops and Selenosis in Enshi

The toxic effects of Se on human health are not commonly observed in natural environments worldwide. However, there are 477 cases of human selenosis reported between 1923 and 1988 in Enshi, China. In 1963, there were 283 people suffering from loss of hair and nail due to selenosis in the region. There are no human selenosis incidents reported in recent years, although the Se toxicity to livestock has been occasionally observed in the villages, showing hoof and hair loss (Fordyce et al. 2000).

To investigate the important factors that control human selenosis, a change in dietary Se intake in the past 50 years (1963–2010) was explored. The local daily dietary Se intake primarily depends on Se concentrations in foodstuff and the amounts of different types of food consumed. Yin and colleagues recently estimated that the food sources of daily dietary Se intake for residents in the Enshi region include cereals (50.7 %), vegetables (17.1 %), meat (15.7 %), tuber (13.5 %), bean (0.2 %), and others (2.8 %).

Maize is the most important cereal crop in Enshi, and the Se content in maize in Enshi had the highest Se concentration of 33.47 mg/kg DW in 1963 (Table 2.5). The average Se concentration in maize significantly decreased to 8.66 mg/kg DW 3 years later in 1966 (Yang et al. 1981a, b). While in the 1980s, the maize Se concentrations varied from 4.17 to 14.07 mg/kg DW (Zheng et al. 1993; Mao et al. 1997). During the early 1990s, maize had stable but low Se concentrations within a

**Table 2.5** Variations on Se contents of food for residents in Enshi during 1963–2010

| Food Source | Se content (mg/kg dry weight) mean (min–max) | Sampling time (year) | References |
|---|---|---|---|
| Maize | 33.47 | 1963 | Mao et al. (1997) |
| | 8.66 (0.5–44.0) | 1966 | Yang et al. (1981a, b) |
| | 14.07 | 1987 | Mao et al. (1997) |
| | 4.17 (0.77–7.57) | 1989 | Zheng et al. (1993) |
| | 6.47 (2.18–10.76) | 1992 | Zhu and Zheng (2001) |
| | 5.95 (4.40–7.50) | 1995 | Yin et al. (1996) |
| | 1.38 (0.182–5.60) | 1996 | Fordyce et al. (2000) |
| | 1.48 (0.07–2.89) | 1999 | Zhu et al. (2008a, b) |
| | 0.37 (0–0.79) | 2010 | This study |
| Rice | 3.96 (0.3–20.2) | 1966 | Yang et al. (1981a, b) |
| | 1.26 (0.83–1.68) | 1995 | Yin et al. (1996) |
| | 1.04 (0.34–1.74) | 2010 | This study |
| Bean | 11.86 (5.0–22.2) | 1966 | Yang et al. (1981a, b) |
| | 0.71 (0.46–1.37) | 2010 | This study |
| Carrot | 11.84 | 1966 | Yang et al. (1981a, b) |
| | 0.23 (0.07–1.60) | 2010 | This study |
| Garlic | 44.80 (8.30; 87.37) | 1966 | Yang et al. (1981a, b) |
| | 0.53 (0.33–1.08) | 2010 | This study |
| Hyacinth bean | 37.23 | 1966 | Yang et al. (1981a, b) |
| | 0.57 (0.11–3.75) | 2010 | This study |
| Chinese cabbage | 36.42 (5.77–72.17) | 1966 | Yang et al. (1981a, b) |
| | 0.72 (0.31–2.73) | 2010 | This study |
| Pumpkin | 33.20 (6.28; 60.02) | 1966 | Yang et al. (1981a, b) |
| | 0.76 (0.31–3.22) | 2010 | This study |
| Eggplant | 38.30 | 1966 | Yang et al. (1981a, b) |
| | 1.04 (0.43–3.21) | 2010 | This study |
| Kidney bean | 28.17 | 1966 | Yang et al. (1981a, b) |
| | 1.57 (0.41–4.14) | 2010 | This study |
| Potato | 9.20 (3.17; 15.13) | 1966 | Yang et al. (1981a, b) |
| | 0.28 (0.04–1.07) | 2010 | This study |

range from 5.95 to 6.47 mg/kg DW (Yin et al. 1996; Zhu and Zheng 2001). However, the Se concentration in maize continued to decrease to 1.38–1.48 mg/kg DW during the late 1990s (Fordyce et al. 2000; Zhu et al. 2008b). Till 2010, the maximum Se concentration of maize was only 0.79 mg/kg DW, while 44.0 mg/kg in maize in 1963. Similar trends were found in rice, bean, carrot, garlic, hyacinth bean, Chinese cabbage, pumpkin, eggplant, kidney bean, and potato (Table 2.5). Concentrations of Se in blood or plasma are common indicators of Se status in the human body (Harrison et al. 1996). However, previous studies revealed that Se concentrations in the muscle and whole blood, red blood cells, blood plasma, hair, and toenails were significantly correlated with each other. In particular, human hair samples have been considered as a good bioindicator for the Se level in the human body (Tan and Huang 1991; Wietecha et al. 2005; Behne et al. 2010).

**Table 2.6** Variations on Se contents of hairs and Se daily dietary intakes for residents in Enshi during 1966–2010

| Se in hair (mg/kg dry weight) mean (min–max) | Se daily dietary intake (µg/d) mean (min–max) | Sampling time (year) | References |
| --- | --- | --- | --- |
| 32.2 (4.1–100) | 4,990 (3,200–6,690) | 1966 | Yang et al. (1981a, b) |
| / | 1,338 | 1985 | Yin et al. (1996) |
| 26.4 (1.832–141) | / | 1996 | Fordyce et al. (2000) |
| 17.49 (9.53–32.82) | 575 (369; 526; 830) | 2010 | This study |

In this study, we collected available data on Se contents of human hairs from Enshi region (Table 2.6). In 1966, the mean Se content in human hair was as high as 32.4 mg/kg DW, which was correspondent to selenosis there. After 30 years, the determined hair Se concentration was lower than that in 1966 and the value was 26.4 mg/kg DW, showing a decrease of 20 %. In this study, we also collected some hair samples from Enshi and the Se content was 17.94 mg/kg DW. Overall, the Se contents in human hair continued to decrease from 32.4 to 17.49 mg/kg DW during the past 45 years, which indicated that the Se level in the human body went down since selenosis occurred in the 1960s.

The adult daily dietary Se intake rates in different countries are compiled in Table 2.7, showing that the Se daily intake varied from 7 to 11 $\mu g\,d^{-1}$ in the Keshan disease area to 600–5,000 $\mu g\,d^{-1}$ in the selenosis areas in Enshi. The recommended dietary allowance (RDA) of Se for humans varied from country, region, age, and sex. In 1980, the estimated safe and adequate daily Se dietary intake for adults was 50–250 $\mu g\,d^{-1}$, and in 1989, the RDA value was established as 77 and 55 $\mu g\,d^{-1}$ of Se for men and women, respectively (Pedrero and Madrid 2009). However, the WHO-recommended-RDA value for adults is 55 $\mu g\,d^{-1}$ for both male and female (National Research Council 2000), and the tolerable upper Se intake level for adults is 400 $\mu g\,d^{-1}$ (Food and Nutrition Board USA Institute of Medicine 2000). Yang et al. (1989) reported that Se homeostasis was disturbed at the Se intake of 750 µg $d^{-1}$ or above, and the symptoms of selenosis occurred at the dietary Se intake level of >910 $\mu g\,d^{-1}$. It was also recommended that 550 µg Se $d^{-1}$ was the maximum safe intake of Se for adults in high Se areas, such as Yutangba.

Based on our collected data, the daily Se intake was as high as 4,990 µg $d^{-1}$ in 1966, much higher than the recommended maximum safe intake by Yang et al. (1989). But the daily Se intake for residents in Enshi continued to decrease from 4,990 µg $d^{-1}$ in 1966 to 1,338 µg $d^{-1}$ in 1985. Although the Se intake in 2010 was significantly lower compared with those in 1966 and 1985, the daily Se intake value still exceeded the recommended maximum safe intake of 550 µg Se $d^{-1}$ (Yang et al. 1989; Yang and Xia 1995), indicating that there may be potential risk for selenosis currently in high Se areas in Enshi.

The Se-enriched coal stone was utilized as fuel materials for cooking and making lime by the villagers in Enshi, and they were also ground into powder as an

**Table 7** A summary on Se intakes from different countries/regions modified from Rayman (2004) and Gao et al. (2011)

| Country/Region | Se intake (µg/person per day) | References |
|---|---|---|
| 1  Keshan disease area (China) | 7 | Yang (1990) |
| Saudi Arabia | 15 | Al-Salehet al. (1997) |
| Czech Republic | 10–25 | Kvicala et al. (1996) |
| Burundi (Africa) | 17 | Benemariya et al. (1993) |
| New Guinea | 20 | Donovan et al. (1992) |
| Nepal | 23 | Moser et al. (1988) |
| China (except Keshan disease area and selenosis) | 26–32 | Chen et al. (2002) |
| Croatia | 27 | Klapec et al. (1998) |
| Egypt | 29 | Reilly (1996) |
| 2  India | 27–48 | Mahalingam et al. (1997) |
| Belgium | 28–61 | Robberecht and Deelstra (1994) |
| Brazil | 28–37 | Maihara et al. (2004) |
| UK | 29–39 | Ministry of Agriculture, Fisheries and Food (1997) |
| France | 29–43 | Lamand et al. (1994) |
| Serbia | 30 | Djujic et al. (1995) |
| Slovenia | 30 | Pokorn et al. (1998) |
| Turkey | 30–36.5 | Giray and Hincal (2004) |
| Poland | 30–40 | Wasowicz et al. (2003) |
| Sweden | 31–38 | Becker (1989) |
| Germany | 35 | Alfthan and Neve (1996) |
| Spain | 35 | Diaz-Alarcon et al. (1996) |
| Portugal | 37 | Reis et al. (1990) |
| Denmark | 38-47 | Danish Governmental Food Agency (1995) |
| Slovakia | 38 | Kadrabova et al. (1998) |
| Greece | 39 | Pappa et al. (2006) |
| Netherlands | 39–67 | Kumpulainen (1993;) van Dokkum (1995) |
| Italy | 43 | Allegrini et al. (1985) |
| Suzhou (China) | 44 | Gao et al. (2011) |
| Austria | 48 | Sima and Pfannhauser (1998) |
| Ireland | 50 | Murphy et al. (2002) |
| 3  Korea | 58 | Choi et al. (2009) |
| Australia | 57–87 | Fardy et al. (1989) |
| New Zealand | 55–80 | Vannoort et al. (2000) |
| Switzerland | 70 | Kumpulainen (1993) |
| Finland | 80 | Hartikainen (2005) |
| 4  Japan | 104–199 | Rayman (2004) |
| USA | 94–134 | Longnecker et al. (1991) |
| Canada | 98–224 | Gissel-Nielsen (1998) |
| 5  Venezuela | 200–350 | Combs and Combs (1986) |
| 6  Selenosis area (China) | 575–4,990 | The present study |

*Note* 1—Se deficiency area; 2—Se low-deficiency area; 3—Se adequate-low area; 4—Se high-adequate area; 5—Se high area; 6—Selenosis area agricultural fertilizer (Zhu et al. 2008b). Moreover, villagers in Enshi also discharged lime onto cropland to improve the soil quality during land clearing for agriculture or cultivation (Yang et al. 1983). These anthropogenic activities accelerated the release and transport of Se from coal stone into the food chain and very likely caused the Se poisoning in Enshi (Zhu et al. 2008b).

## 2.8 Selenium-Biofortified Agricultural Products in Enshi

Several earlier clinical trials have suggested that some organic forms of Se could lower the risk of certain types of cancer (Clark et al. 1996; Reid et al. 2008; Wallace et al. 2009).

Se daily intake data from the world are compiled in Table 2.7, which displays the intake of Se varying considerably between countries/regions. Keshan disease area (China), Saudi Arabia, Czech Republic, Burundi (Africa), New Guinea, Nepal, China (except KD and selenosis), Croatia, and Egypt were identified as Se-deficiency countries/regions because the levels of Se daily intake were below 30 μg/d; India, Belgium, Brazil, UK, France, Serbia, Slovenia, Turkey, Poland, Sweden, Germany, Spain, Portugal, Denmark, Slovakia, Greece, Netherlands, Italy, Suzhou (China), Austria, Ireland were identified as Se-low to Se deficiency area because the levels of Se daily intake were below the WHO recommended amount, 55 μg/d; Korea, Australia, New Zealand, Switzerland, and Finland were identified as Se-adequate to Se-low areas because of the levels of Se daily intake were in a range of 55–100 μg/d; Japan, USA, and Canada were recognized as Se-high to Se-adequate countries with the Se daily intake of 100–200 μg/d. It is quite a high level of Se intake in Venezuela with 200–350 μg/d; if the residents took more than 550 μg/d Se, it would cause selenosis symptoms, such as in Enshi, China. Overall, there are about 76 % (28/35) of countries located in Se-low areas with the Se daily intake level less than 55 μg/d. Especially in China, the Se daily intake varied considerably from toxic in Enshi, through low in Suzhou, to deficient in Keshan disease areas.

Soil Se distribution varied significantly in the world. More than 40 countries lack Se resources, while about 80 % of the world's total reserves of Se are located in Chile, the United States, Canada, China, Zambia, Zaire, Peru, Philippines, Australia, and Papua New Guinea (Liu et al. 2010). Although China is ranked fourth in Se reserves worldwide, after Canada, the United States, and Belgium, Se-deficiency occurs in a geographic low-Se belt stretching from Heilonjiang Province in the northeast to Yunnan Province in the southwest, affecting 71.2 % of Chinese land (Zhu et al. 2009). Therefore, Se food supplement is needed for many Chinese people. Till date, plant-based Se intake has been the only means for humans and animals in Se-deficient areas. Wheat, rice, and vegetables are usually Se-biofortificated to provide organic and safe Se compounds (Zhu et al. 2009; Liu et al. 2010).

## 2.8.1 Selenium Biofortification Strategy

Biofortification is a biological strategy, which aims to increase micronutrient contents in the edible parts of plants, animals, or microorganisms, via breeding or the use of biotechnology. It is considered to be a safe and effective way to alleviate micronutrient malnutrition in many micronutrient deficient/low areas or countries (Nestel et al. 2006; Mayer et al. 2008; Zhao and McGrath 2009). Generally, plant-based biofortification is the most effective and worldwide used strategy, especially on staple crops, because it is the best solution for improving the lack of nutritional trace elements in the world (White and Broadley 2009).

However, Se is not an essential micronutrition for higher plants, and Se will be transported via S-transportation pathway into plant tissues (Terry et al. 2000). In fact, the ability to absorb and accumulate Se is different in different plant species. Therefore, it is important to select special plant species which can accumulate Se in their edible parts for biofortification. Then, the biofortification strategies are used on these selected plants to increase the Se concentrations in the edible parts, which can be consumed by populations in Se malnutrition status. Furthermore, plants accumulating Se are useful as a "Se-delivery system" to supplement Se in the mammalian diet in many Se-deficient countries or areas, and these Se-bio-fortificated meat could be another important source for dietary Se intake. In addition, the un-edible parts of biofortified plants and the excrements of fortified animals could also be used as (organic) Se-enriched fertilizers for staple crops.

There are two strategies currently for Se-biofortification, agronomic approaches and genetic approaches.

(1) Agronomic biofortification strategies

Agronomic biofortification strategies are based on application of mineral fertilizers to improve the solubilization and mobilization of Se in the soil (White and Broadley 2009). The different forms of Se supplied for biofortification could be different in the Se accumulation of higher plants. Selenate is transported much more easily than selenite, or organic Se, and plant leaves could accumulate substantial amounts of selenate but much less selenite or SeMet (De Souza et al. 1998; Zayed et al. 1999). In addition, the mixture of organic acids with Se-mineral fertilizers were used to chelate Se, which could obviously improve the acquisition of Se and elevate the utilization efficiency of Se fertilizers (Morgan et al. 2005; Lynch 2007). It is also an effective approach to develop a more extensive root system, with longer, thinner roots with more root hairs, and by proliferating lateral roots in mineral-rich patches (White and Broadley 2005; Lynch 2007; Kirkby and Johnston 2008; White and Hammond 2008). Moreover, the rhizosphere microorganisms played an important role in phytoavailability of Se by plants (Morgan et al. 2005; Lynch 2007; Kirkby and Johnston 2008). It should be pointed out that the agronomic Se-biofortification strategies to increase crop Se contents by using inorganic Se fertilizers were very successful in Finland and New Zealand (Lyons et al. 2003; Hartikainen 2005). Clearly, it is promising to use the Se-enriched plants, crops, or agricultural products grown on naturally seleniferous soils, for example, in Enshi of China, as a natural Se supplement for people in areas with inadequate soil Se concentrations (Terry et al. 2000).

(2) Genetic engineering for biofortification

Genetic biofortification strategies are based on genetic variations or transgenic technology to increase abilities to acquire the objective micronutrient elements and accumulate them in edible parts of plants (White and Broadley 2009). Additionally, it is known that so-called "promoter" substances, such as ascorbate, $\beta$-carotene, and cysteine-rich polypeptides, could accelerate the absorption of micronutrient elements in plants, and it is possible to increase the concentrations of mineral elements in plants by increasing the contents of "promoter" substances in genetic ways. It is the reverse with "antinutrient" substances, such as oxalate, polyphenolics, or phytate (White and Broadley 2009). There is genetic variation in the concentrations of mineral elements in the grains of most cereal species. Some researches indicate that concentrations of Fe and Zn in cereal grain vary 1.5- to 4-fold among genotypes depending on the genetic diversity of the material tested (Cakmak 2008; Tiwari et al. 2008). Generally, the Se levels in different plants are as follows: brassica > bean > cereal (Liu et al. 2010). As for transgenic approaches, the selenocysteine methyltransferase gene of *Astragalus bisulcatus* (two-grooved poison vetch) was introduced into *Arabidopsis thaliana* (Thale cress) to overexpress Se-methylselenocysteine and $\gamma$-glutamylmethylselenocysteine in shoots (Ellis et al. 2004; Sors et al. 2005; Pilon-Smits and LeDuc 2009).

## 2.8.2 Selenium Biofortification in China

Considering that there are so many Se-deficient regions in the world, it is promising to take advantage of Se-enriched plants and crops in Enshi as a natural and green Se resource for animals and human beings.

One option is to add the Se-enriched plants in Enshi to soils in other Se-deficient areas as a source of organic Se fertilizer supporting forage crops. Proper amounts of this organic fertilizer can improve the Se status in the local soil as well as provide the crops with Se and other nutrition. Second, it is a good solution to use Se-enriched plant materials in Enshi as forage for animals in other Se-deficiency areas. Third, the Se-enriched staple crops in Enshi could be regarded as naturally Se-biofortified products, and those Se-enriched products could be consumed by populations in Se-deficiency areas as Se food supplement sources. The local business in Enshi has developed some Se-enriched products, such as tea, rice, maize, herb, and drinks.

The development of Se-biofortification has been ongoing for decades in China (Yang et al. 2007). Generally, Se biofortification approaches in China can be divided into three different categories:

(1) Selective Se-accumulated crop species

The black rice Jinlong No.1, cultivated by Jilin Academy of Agricultural Sciences, could accumulate Se with a content of 6.5 µg/g DW. Jiangsu Academy of Agricultural Sciences cultivated an Se-enriched rice species, named Longqing No.4, which was optimized from Suzi No.4 in Yunnan province. Shanxi Academy of Agricultural Sciences bred a new black wheat variety with Se concentration 112.8 % higher than the ordinary one. Furthermore, the selected Se-accumulated species, e.g., black rice, red rice, could significantly increase the content of Se in the edible parts. It is possible to mutagenize the Se-related genes in *Arabidopsis thaliana* to improve the efficiency of breeding Se-enriched crops at molecular level (Liu et al. 2010).

(2) Foliar application of Se fertilizer

Foliar spray with Se fertilizer is a practical way to improve the Se content of staple crops in China, and it played an important role in producing Se-enriched foods. Under the optimal application condition, Se contents of rice could be significantly increased by 194 % and reached over 120 µg/kg (DW) without reducing grain yields and protein/ash contents (Fang et al. 2008). Chen et al. (2002) also found that the Se contents of rice were significantly increased to 0.471–0.640 µg/g by foliar application of Se-fertilizer at a rate of 20 g Se/ha in the forms of sodium selenite and sodium selenate. At present, Se-enriched rice is available in the market and contributes significantly to consumers by improving their Se dietary intake since rice is one of the major staple foods in China. Tea is another popular Se biofortified product in China. Besides the Se contents of tea leaves being increased, the number of sprouts, the yield, the amino acid contents, the vitamin C contents as well as the sweetness and aroma of tea leaves could be significantly increased because of the implementation of the Se biofortification strategy (Hu et al. 2003).

(3) Application of soil Se fertilizers

This approach is to apply Se fertilizers around plant root zone to increase the total Se content and bioavailable Se in the rhizosphere environment. Compared with natural biofortification and foliar spray approaches, the application of soil Se fertilizers has the following advantages: (1) it breaks down the geological limitations for Se biofortification, compared with the natural biofortification in seleniferous areas; (2) the Se chemical forms and contents in Se-biofortified products would be much safer than those via foliar spray; and (3) it could largely reduce the deviation of Se contents in the biofortified products to ensure high quality on Se-enriched products in future.

Generally, fruits and vegetables in China contain less than 3 µg/kg Se (wet weight) while rice less than 50 µg/kg, and tea less than 250 µg/kg (Yin and Li 2011). However, the use of soil Se fertilizers could improve the Se contents in the products by several hundred times, and it was performed on various cereals, fruits, and vegetables (Liu et al. 2010; Yin and Li 2011). In recent years, the Se fertilizer application strategy was commonly used in Chinese agricultural production and produced safe and green Se-enriched foods in the market, such as fruits, vegetables, rice, and tea. Indeed, the novel concept of functional agriculture had been adopted by Chinese scientists and it has received more and more recognition from growers to consumers (Zhao and Huang 2010).

## 2.9 Summary and Outlooks

Selenium is an essential mineral nutrient for humans and animals. Selenium is needed for the formation of several proteins such as glutathione peroxide and thioredoxin reductase. However, the gap between the beneficial and harmful levels of Se is quite narrow. The Keshan disease and the Kaschin-Beck disease caused by Se deficiency occurred in Heilongjiang, China, with a daily Se intake less than 11 µg/d and the loss of hair and nail caused by Se poisoning occurred in Enshi, central China, with a daily Se intake more than 575 µg/d. Therefore, the concurrent endemic diseases of Se-deficiency and selenosis that happened in China indicated the greatly uneven distribution of Se resources in China.

Although the Se concentration in resident foods and the daily Se intake decreased significantly from 1963 to 2010 in Enshi, the present daily Se intake (575 µg/d) is still above the recommended maximum safe intake with 550 µg/d, which indicates there may be potential risk for selenosis in Enshi. Moreover, the total soil Se content in Enshi concentrated in a range from 20 to 60 mg/kg DW which was approximately 150–500 times greater than the average Se content (0.125 mg/kg DW) in Se-deficient areas and approximately 50–150 times greater than that (0.40 mg/kg DW) in Se-riches areas in China, respectively.

In contrast, there are about 76 % countries located in Se-deficiency areas with the Se daily intake level less than 55 µg/d for adults. Especially in China, Se-deficiency occurs in a geographic low-Se belt stretching from Heilonjiang Province in the northeast to Yunnan Province in the southwest, covering about 70 % of Chinese land.

Therefore, it is promising to take Se-biofortification naturally in Enshi. Se-enriched plants or crops in Enshi could be taken as a source of Se-organic fertilizer to increase the Se contents of staple crops, or as a source of Se-organic forage to support the Se-deficiency livestock in Se-deficient areas. Se-enriched crops, such as rice, maize, could be consumed by the population as a safe Se-supplement in Se deficiency areas. Furthermore, an Se-hyperaccumulating plant, *Adenocaulon himalaicum,* could be planted widely in Enshi to gain high-Se materials, and it could also be biofortified in Se deficiency areas as a selective species.

# References

Alfthan G, Neve J (1996) Se intakes and plasma Se levels in various populations. In: Kumpulainen J, Salonen J (eds) Natural antioxidants and food quality in atherosclerosis and cancer prevention. Royal Society of Chemistry, Cambridge, pp 161–167

Allegrini M, Lanzola E, Gallorini M (1985) Dietary Se intake in a coronary heart disease study in Northern Italy. Nutr Res Suppl 1:398–402

Al-Saleh I, Al-Doush I, Faris R (1997) Se levels in breast milk and cow's milk: a preliminary report from Saudi Arabia. J Environ Pathol Toxicol Oncol 16:41–46

Becker W (1989) Food habits and nutrient intake in Sweden 1989. Swedish National Food Administration, Uppsala

Behne D, Alber D, Kyriakopoulos A (2010) Long-term Se supplementation of humans: Se status and relationships between Se concentrations in skeletal muscle and indicator materials. J Trace Elem Med Biol 24:99–105

Beilstein MA et al (1991) Chemical forms of Se in corn and rice grown in a high Se area of China. Biomed Environ Sci 4:392–398

Bell PF, Parker DR, Page AL (1992) Contrasting selenate sulfate interactions in Se accumulating and nonaccumulating plant species. Soil Sci Soc Am 56:1818–1824

Benemariya H, Robberecht H, Deelstra H (1993) Daily dietary intake of copper, zinc and Se by different population groups in Burundi, Africa. Sci Total Environ 136:49–76

Cakmak I (2008) Enrichment of cereal grains with zinc: agronomic or genetic biofortification? Plant Soil 302:1–17

Chen L, Yang F, Xu J, Hu Y, Hu Q, Zhang Y, Pan GJ (2002) Determination of Se concentration of rice in China and effect of fertilization of selenite and selenate on Se content of rice. J Agric Food Chem 50(18):5128–5130

Choi Y, Kim J, Lee H, Kim C, Hwang IK, Park HK (2009) Se content in representative Korean foods. J Food Compos Anal 22:117–122

Clark LC, Combs GF, Turnbull BW, Slate EH, Chalker DK, Chow J, Davis LS, Glover RA, Graham GF, Gross EG, Krongrad A, Lesher JL, Park HK, Sanders BB, Smith CL, Taylor JR (1996) Effects of Se supplementation for cancer prevention in patients with carcinoma of the skin a randomized controlled trial—a randomized controlled trial. J Am Med Assoc 276:1957–1963

Combs GF Jr, Combs SB (1986) The biological availability of Se in foods and feeds. In the role of Se in nutrition. Academic Press, New York, pp 127–177

Danish Governmental Food Agency (1995) Food Habits of Danes 1995, Main Results. Levnedsmiddelstyrelsen, Soeborg

De Souza MP, Pilon-Smits EAH, Lytle CM, Hwang S, Tai J et al (1998) Rate-limiting steps in Se assimilation and volatilization by Indian mustard. Plant Physiol 117:1487–1494

Diaz-Alarcon JP, Navarro-Alarcon M, Lopez-Garcia de la Serrana H, Lopez-Martinez MC (1996) Determination of Se in meat products by hydride generation atomic absorption spectrophotometry—Se levels in meat, organ meats and sausages in Spain. J Agric Food Chem 44:1494–1497

Djujic I, Djujic B, Trajkovic L (1995) Dietary intake of Se in Serbia: results for 1991. Naucni Skupovi (Srpska Akademija Nauka I Umetnosti). Odeljenje Prirodno-Matematickih Nauka 6:81–87

Donovan UM, Gibson RS, Ferguson EL, Ounpuu S, Heywood P (1992) Se intakes of children from Malawi and Papua New Guinea consuming plant-based diets. J Trace Elem Electrolytes Health Dis 6:39–43

Ellis DR, Salt DE (2003) Plants, Se and human health. Curr Opin Plant Biol 6:273–279

Ellis DR, Sors TG, Brunk DG et al (2004) Production of Se-methylselenocysteine in transgenic plants expressing selenocysteine methyltransferase. BMC Plant Biol 4:1–11

Fang Y et al (2008) Effect of foliar application of zinc, selenium, and iron fertilizers on nutrients concentration and yield of rice grain in China. J Agric Food Chem 56(6):2079–2084

Fardy JJ, McOrist GD, Farrar YJ (1989) The determination of Se in the Australian diet using neutron activation analysis. J Radioanal Nucl Chem 133:391–396

Feist LJ, Parker DR (2001) Ecotypic variation in Se accumulation among populations of Stanleya pinnata. New Phytol 149:61–69

Food and Nutrition Board USA Institute of Medicine (2000) Dietary references intakes for vitamin C, vitamin E, Se and carotenoids. National Academy Press, Washington, pp 284–324

Fordyce FM, Zhang G, Green K, Liu X (2000) Soil, grain and water chemistry in relation to human Se-responsive disease in Enshi District, China. Appl Geochem 15:117–132

Freeman JL, Quinn CF, Marcus MA, Fakra S, Pilon-Smits EAH (2006) Se tolerant diamondback moth disarms hyperaccumulator plant defense. Curr Biol 16:2181–2192

Galeas ML et al (2007) Seasonal fluctuations of Se and sulfur accumulation in Se hyperaccumulators and related nonaccumulators. New Phytol 173:517–525

Gao J, Liu Y, Huang Y, Lin ZQ, Banuelos GS, Lam MHW, Yin XB (2011) Daily Se intake in a moderate Se deficiency area of Suzhou, China. Food Chem 126:1088–1093

Giray B, Hincal F (2004) Se status in Turkey. J Radioanal Nucl Chem 259:447–451

Gissel-Nielsen G (1998) Effects of Se supplementation of field crops. In: FrankenbergerJr WT, Engberg RA (eds) Environmental chemistry of Se. Marcel Dekker, New York, pp 99–112

Harrison I, Littlejohn D, Fell GS (1996) Distribution of Se in human blood plasma and serum. Analyst 121:189–194

Hartikainen H (2005) Biogeochemistry of Se and its impact on food chain quality and human health. J Trace Elem Med Biol 18:309–318

Hu Q, Xu J, Pang GJ (2003) Effect of Se on increasing the antioxidant activity of tea leaves harvested during the early spring tea producing season. J Agric Food Chem 51(11):3379–3381

Kadrabova J, Madaric A, Ginter E (1998) Determination of the daily Se intake in Slovakia. Biol Trace Elem Res 61:277–286

Kahakachchi C et al (2004) Chromatographic speciation of anionic and neutral Se compounds in Se-accumulating Brassica juncea (Indian mustard) and in selenized yeast. J Chromatogr A 1054:303–312

Kirkby EA, Johnston AE (2008) Soil and fertilizer phosphorus in relation to crop nutrition. In: Hammond JP, White PJ (eds) The ecophysiology of plant-phosphorus interactions. Springer, Dordrecht, pp 177–223

Klapec T, Mandic ML, Grigic J, Primorac L, Ikic M, Lovric T, Grigic Z, Herceg Z (1998) Daily dietary intake of Se in eastern Croatia. Sci Total Environ 217:127–136

Kumpulainen JT (1993) Se in foods and diets of selected countries. J Trace Elem Electrolytes Health Dis 7:107–108

Kvicala J, Zamrazil V, Jiranek V (1996) Se deficient status of the inhabitants of South Moravia. In: Kumpulainen J, Salonen J (eds) Natural antioxidants and food quality in atherosclerosis and cancer prevention. Royal Society of Chemistry, Cambridge, pp 177–187

Lamand M, Tressol JC, Bellanger J (1994) The mineral and trace element composition in French food items and intake levels in France. J Trace Elem Electrolytes Health Dis 8:195–202

Li SH, Xiao TF, Zheng BS (2011) Medical geology of arsenic, Se and thallium in China. Sci Total Environ. doi:10.1016/j.scitotenv.2011.02.040

Liu Y, Li F, Yin XB, Lin ZQ (2010) Plant-based biofortification: from phytoremediation to Se-enriched agriculture products. In: Green chemistry for environmental sustainability. CRC press, Boca Raton

Longnecker MP, Taylor PR, Levander OA, Howe SM, Veillon C, Mcadam PA et al (1991) Se in diet, blood, and toenails in relation to human health in a seleniferous area. Am J Clin Nutr 53:1288–1294

Lynch JP (2007) Roots of the second green revolution. Aust J Bot 55:493–512

Lyons G, Stangoulis J, Graham R (2003) High-Se wheat: biofortification for better health. Nutr Res Rev 16:45–60

Mahalingam TR, Vijayalakshni S, Prabhu RK et al (1997) Studies on some trace and minor elements in blood. A survey of the Kalpakkan (India) population. Part III: Studies on dietary intake and its correlation to blood levels. Biol Trace Elem Res 57:223–238

Maihara VA, Gonzaga IB, Silva VL, Favaro DIT, Vasconcellos MBA, Cozzolino SMF (2004) Daily dietary Se intake of selected Brazilian population groups. J Radioanal Nucl Chem 259:465–468

Mao J, Xing B (1999) Fractionation and distribution of Se in soils. Commun Soil Sci Plant Anal 30(17/18):2437–2447

Mao DJ, Su HC, Yan LR (1990) An epidemiologic investigation on Se poisoning in southwestern Hubei Province. Chin J Endemiol 9:311–314 (In Chinese with English Abstract)

Mao DJ, Zheng BS, Su HC (1997) The medical geography characteristics of Se-poisoning in Yutangba. Endem Dis Bull 12:59–61 (In Chinese with English Abstract)

Mayer JE, Pfeiffer WH, Beyer P (2008) Biofortified crops to alleviate micronutrient malnutrition. Curr Opin Plant Biol 11:166–170

Mei ZQ (1985) Summary on two Se-rich areas of China. Chin J Endem 4:379–385 (In Chinese with English Abstract)

Ministry of Agriculture, Fisheries and Food (1997) Ministry of Agriculture, Fisheries and Food, October 1997, food surveillance information sheet, no. 126. Dietary intake of Se. Joint Food Safety and Standards Group,London

Morgan JAW, Bending GD, White PJ (2005) Biological costs and benefits to plant-microbe interactions in the rhizosphere. J Exp Bot 56:1729–1739

Moser PB, Reynolds RD, Acharya S, Howard MP, Andon MB, Lewis SA (1988) Copper, iron, zinc, and Se dietary intake and status of Nepalese lactating women and their breast-fed infants. Am J Clin Nutr 47:729–734

Murphy J, Hannon EM, Kiely M, Flynn A, Cashman KD (2002) Se intakes in 18–64-y-old Irish adults. Eur J Clin Nutr 56:402–408

National Research Council (2000) Dietary Reference Intakes (DRI). National Academy Press, Washington, pp 284–319

Nestel P, Bouis HE, Meenakshi JV, Pfeiffer W (2006) Biofortification of staple food crops. J Nutr 136:1064–1067

Pappa EC, Pappas AC, Surai PF (2006) Se content in selected foods from the Greek market and estimation of the daily intake. Sci Total Environ 372:100–108

Pedrero Z, Madrid Y (2009) Novel approaches for Se speciation in foodstuffs and biological specimens: a review. Anal Chim Acta 634:135–152

Pilon-Smits EAH, LeDuc DL (2009) Phytoremediation of Se using transgenic plants. Curr Opin Biotechnol 20:207–212

Pilon-Smits EAH, Quinn CF (2010) Se metabolism in plants. In: Hell R, Mendel PR (eds) Cell biology of metals and nutrients. Plant cell monographs vol 17. Springer-Verlag, Berlin doi:10.1007/978-3-642-10613-2-10

Pokorn D, Stibilj V, Gregoric B, Dermelj M, Stupar J (1998) Elemental composition (Ca, Mg, Mn, Cu, Cr, Zn, Se and I) of daily diet samples from some old people's homes in Slovenia. J Food Compos Anal 11:47–53

Presser TS, Marc AS, Walton HL (1994) Bioaccumulation of Se from natural geologic sources in western states and its potential consequences. Environ Manag 18:423–436

Rayman MP (2004) The use of high-Se yeast to raise Se status: how does it measure up? Br J Nutr 92:557–573

Reid ME, Duffield-Lillico AJ, Slate E, Natarajan N, Turnbull B, Jacobs E, Combs GF, Alberts DS, Clark LC, Marshall JR (2008) The nutritional prevention of cancer: 400 mcg per day Se treatment. Nutr Cancer 60:155–163

Reilly C (1996) Se in food and health. Blackie Academic and Professional, London

Reis MF, Holzbecher J, Martinho E, Chatt A (1990) Determination of Se in duplicate diets of residents of Pinhel, Portugal, by neutron activation. Biol Trace Elem Res 26–27:629–635

Robberecht HJ, Deelstra HA (1994) Factors influencing blood Se concentrations: a literature review. J Trace Elem Electrolytes Health Dis 8:129–143

Sharmasarkar S, Vance GF (1995) Fractional partitioning for assessing solid phase speciation and geochemical transformations of soil Se. Soil Sci 160(1):43–55

Sima A, Pfannhauser W (1998) Se levels in foods produced in Austria. In: Anke M (ed) Mengen-Spurenelem, Arbeitstag 18th. Verlag Harald Schubert, Leipzig, pp 197–204

Sors TG, Ellis DR, Na GN et al (2005) Analysis of Sulfur and Se assimilation in *Astragalus* plants with varying capacities to accumulate Se. Plant J 42:785–797

Stadlober M et al (2001) Effects of selenate supplemented fertilisation on the Se level of cereals—identification and quantification of Se compounds by HPLC-ICP-MS. Food Chem 73:357–366

Tan JA, Huang YJ (1991) Se in geo-ecosystem and its relation to endemic diseases in China. Water Air Soil Pollut 57(58):59–65

Tan JA, Zhu WY, Wang WY et al (2002) Se in soil and endemic diseases in China. Sci Total Environ 284:227–235

Terry N, Zayed AM, de Souza MP, Tarun AS (2000) Se in higher plants. Annu Rev Plant Physiol Plant Mol Biol 51:401–432

Tiwari VK, Rawat N, Neelam K, Randhawa GS, Singh K, Chhuneja P, Dhaliwal HS (2008) Development of *Triticum turgidum subsp. durum—Aegilops longissima* amphiploids with high iron and zinc content through unreduced gamete formation in F1 hybrids. Genome 51:757–766

van Dokkum W (1995) The intake of selected minerals and trace elements in European countries. Nutr Res Rev 8:271–302

Vannoort R, Cressey P, Silvers K (2000) 1997/1998 New Zealand total diet survey. Part 2: elements. Ministry of Health, Wellington

Wallace K, Kelsey KT, Schned A, Morris JS, Andrew AS, Karagas MR (2009) Se and risk of bladder cancer: a population-based case-control study. Cancer Prev Res 2:70–73

Wang ZJ, Gao YX (2001) Biogeochemical cycling of Se in Chinese environments. Appl Geochem 16:1345–1351

Wasowicz W, Gromadzinska J, Rydzynski K, Tomczak J (2003) Se status of low-Se area residents: polish experience. Toxicol Lett 137:95–101

White PJ, Broadley MR (2005) Biofortifying crops with essential mineral elements. Trends Plant Sci 10:586–593

White PJ, Hammond JP (2008) Phosphorus nutrition of terrestrial plants. In: White PJ, Hammond JP (eds) The ecophysiology of plant-phosphorus interactions. Springer, Dordrecht, pp 51–81

White PJ, Broadley MR (2009) Biofortification of crops with seven mineral elements often lacking in human diets—iron, zinc, copper, calcium, magnesium, Se and iodine. New Phytol 182:49–84

WHO (1987) Environmental health criteria. Se environmental health criteria, vol 58. World Health Organization, Geneva, p 1–110

Wietecha R, Koscielniak P, Lech T, Kielar T (2005) Simple method for simultaneous determination of Se and arsenic in human hair by means of atomic fluorescence spectrometry with hydride generation technique. Microchim Acta 149:137–144

Wilber CG (1980) Toxicology of Se: a review. Clin Toxicol 17:171–230

Wu L, Mantgem PJV, Guo X (1996) Effects of forage plant and field legume species on soil Se redistribution, leaching and bioextraction in soils contaminated by agricultural drain water sediment. Arch Environ Contam Toxicol 31:329–338

Yang GQ, Xia YM (1995) Studies on human dietary requirements and safe range of dietary intakes of Se in China and their application to the prevention of related endemic diseases. Biomed Environ Sci 8:187–201

Yang GQ, Wang SZ, Zhou RH, Sun SZ, Man RE (1981a) Research on the etiology of an endemic disease characterized by loss of nails and hair in Enshi county. J Chin Acad Med 3(Suppl. 2):1–6 (In Chinese with English Abstract)

Yang GQ, Wang SZ, Zhou RH, Sun SZ, Man RE, Li Shensi, Li Zhenbo, Zhao Xianrong, Liu Biansheng, Shen Changling, Mei Huangyou, Wang Xiangjiu (1981b) Investigation on loss of hair and nail of unkown etiology-endemic selenosis. Acta Academiae Medicinae Sinicae 3(Suppl 2):1–6 (In Chinese with English Abstract)

Yang GQ, Wang SZ, Zhou RH, Sun SZ (1983) Endemic Se intoxication of humans in China. Am J Clin Nutr 7:872–881

Yang GQ, Yin S, Zhou R, Gu L, Yan B, Liu Y, Liu Y (1989) Studies of safe maximal daily Se intake in a seleniferous area in China. II. Relation between Se intake and the manifestation of clinical signs and certain biochemical alterations in blood and urine. J Trace Elem Electrolytes Health Dis 3:123–130

Yang GQ (1990) Applied nutrition manual. Science and Technology Press, Tianjin, p 89–107 (In Chinese)

Yang XE et al (2007) Improving human micronutrient nutrition through biofortification in the soil-plant system: China as a case study. Environ Geochem Health 29(5):413–428

Yin XB, Li F (2011) The standardization in Se biofortification. In: Banuelos GS, Lin ZQ, Yin XB, Duan N (eds) Se global perspectives of impacts on human, animals and the environment. University of Science and Technology of China Press, Hefei, pp 113–114

Yin SA, Zhou RH, Yang GQ, Man RG, Xu JR, Yan BW (1996) Effects of milling and cooking on Se content of foods from the sites with different Se levels. Acta Nutrimenta Sinica 18(3):367–370

Yu RY (1993) Preliminary analysis of the geological characteristics and origin of Shuanghe Se ore beds. Hubei Geol 7(1):50–56 (In Chinese with English Abstract)

Zhang GD, Ge XL, Zhang QL et al (1998) Se geological and geochemical environmental background in Enshi, Hubei. Acta Geoscientia Sinica—Bulletin of the Chinese Academy of Geologicla. Science 19(1):59–67 (In Chinese with English Abstract)

Zayed A, Lytle CM, Terry N (1999) Accumulation and volatilization of different chemical species of Se by plants. Planta 206:284–292

Zhang YL, Pan XG, Hu QH, Qiu DS, Chu QH (2002) Se fractionation and bioavailability in some low-Se soils of central Jiansu Province. Plant Nutr Fertil Sci 8(3):355–359 (In Chinese with English Abstract)

Zhao FJ, McGrath SP (2009) Biofortification and phytoremediation. Curr Opin Plant Biol 12:373–380

Zhao QG, Huang JK (2010) Agricultural science and technology in China: a roadmap to 2050. Science Press, Berlin, pp 100–126

Zhao CY, Ren JH, Xue CZ, Lin E (2005) Study on the relationship between soil Se and plant Se uptake. Plant Soil 277:197–206

Zheng BS, Yan LR, Mao DJ, Thornton I (1993) The Se resource in southwestern Hubei province, China, and its exploitation strategy. J Nat Resour 8:204–212 (In Chinese with English Abstract)

Zhu JM, Zheng BS (2001) Distribution of Se in mini-landscape of Yutangba, Enshi, Hubei Province China. Appl Geochem 16:1333–1344

Zhu JM, Qin HB, Li L, Feng ZG, Su HC (2008a) Fractionation of Se in high-Se soils from Yutangba, Enshi Hubei. Acta Scientiae Circumstantiae 28(4):772–777

Zhu JM, Wang N, Li SH et al (2008b) Distribution and transport of Se in Yutangba, China: impact of human activities. Sci Total Environ 392:252–261

Zhu YG, Pilon-Smits EAH, Zhao FJ, Williams PN, Meharg AA (2009) Se in higher plants: understanding mechanisms for biofortification and phytoremediation. Trends Plant Sci 14(8):436–442

# Chapter 3
# Phytoremediation of Zinc-Contaminated Soil and Zinc-Biofortification for Human Nutrition

Li Zhao, Linxi Yuan, Zhangmin Wang, Tianyu Lei and Xuebin Yin

**Abstract** Phytoremediation of Zinc (Zn) contaminated soil is to remove Zn from soil using Zn-tolerant plants or hyperaccumulators with the assistance of agricultural or biological technologies. Zn biofortification is to increase the natural content of Zn with high bioavailability in staple food crops to provide Zn supplement sources for humans. Researches on understanding the physiological mechanisms of Zn uptake, distribution, storage, and metabolism by plants indicate that the two applications share certain limiting physiological processes. In this chapter, the physiological processes of Zn in plants are introduced and certain regulatory mechanisms are reviewed. Issues related to existing strategies, bottlenecks, and potential improvements on Zn-phytoremediation and Zn-biofortification will also be discussed. Though much remains to be elucidated, the combination of high efficiency of Zn accumulation in plants for phytoremediation and favorable Zn-accumulation in the edible part of staple crops for biofortification appear to be a worthwhile and promising attempt.

**Keywords** Zinc · Zinc phytoremediation · Zinc biofortification · Physiological processes

L. Zhao · L. Yuan · Z. Wang · T. Lei · X. Yin (✉)
Advanced Laboratory for Selenium and Human Health, Suzhou Institute for Advanced Study, University of Science and Technology of China, Suzhou 215123, Jiangsu, China
e-mail: xbyin@ustc.edu.cn

X. Yin
School of Earth and Space Science, University of Science and Technology of China (USTC), Hefei 230026, Anhui, China

X. Yin and L. Yuan (eds.), *Phytoremediation and Biofortification*,
SpringerBriefs in Green Chemistry for Sustainability,
DOI: 10.1007/978-94-007-1439-7_3, © The Author(s) 2012

## 3.1 Zinc: An Overview

Zinc (Zn) is the twenty-fourth most abundant element on the Earth's crust with a content of 75 mg kg$^{-1}$. Soil contains 5–770 mg kg$^{-1}$ of Zn with an average of 64 mg kg$^{-1}$, seawater has 30 μg Zn L$^{-1}$, and the atmosphere contains 0.1–4 μg Zn m$^{-3}$ (Emsley 2001). Zn in the environment mainly exists in the state of Zn-sulfide and Zn-oxides, and easily associates with many other elements, such as lead (Pb), copper (Cu), cadmium (Cd) to form mineral associations. Zn is a transition metal with atomic number of 30 and has five stable isotopes: $^{64}$Zn (48.63 %), $^{66}$Zn (27.90 %), $^{67}$Zn (4.90 %), $^{68}$Zn (18.75 %), and $^{70}$Zn (0.62 %). $^{65}$Zn is radioactive, with a half-life time of 244.26 days, and it is frequently used as a Zn radiotracer in plants. Zn commonly presents in oxidation states of +1 or +2 in the environment (Brady et al. 1983). $Zn^{2+}$ participates in biological or chemical reactions. In solution, Zn exists in the +2 oxidation state and is redox-stable under physiological conditions for a complete d-shell of electrons (Broadley 2007). Zn tends to form strong covalent bonds with O-, N-, and S- donors and thus forms numerous stable complexes (Greenwood and Earnshaw 1997). Tetracoordinated and hexacoordinated Zn complexes are the most common types, though penta-coordinated complexes also exist (Holleman et al. 1985).

Zn is an essential trace element for plants (Broadley et al. 2007), animals (Prasad 2008), and microorganisms (Sugarman 1983). It is required in a large number of proteins in organisms and is the only metal presenting in all classes of enzymes. In proteins, Zn ions are often coordinated to the side chains of amino acids, such as cysteine and histidine, aspartic acid, and glutamic acid. In organisms, three main functional sites of Zn are recognized, and the function varies with the geometry and characteristics of $Zn^{2+}$-ligand bonding: structural, catalytic, and cocatalytic (Auld 2001; Maret 2005). Since Zn has no oxidant properties like iron and copper, it exists almost entirely in divalent state, making it easily incorporated into the biological system and safely transports both extra- and intracellularly (Hambidge and Krebs 2007).

Zn is responsible for the normal expression of more than 20 physiological functions in organisms, including immune function, protein synthesis, wound healing, DNA synthesis, and cell division. A large number of proteins in biological systems need Zn to maintain their structural stability and transcription factors. Protection against infections and diseases is related to gene regulation and expression under stress conditions in which Zn is required (Prasad 2010). Zn is also a critical element required for detoxification of highly aggressive free radicals and for structural and functional integrity of biological membranes (Cakmak 2000). Zn supports normal growth and development during pregnancy, childhood, and adolescence. It is also required for proper sense of taste and smell. Zn deficiency is among the major malnutrition in humans and has led to severe diseases, cellular disturbances and impairments, and even large mortality especially in infants and young children.

Although Zn is an essential element for life, excess Zn can be harmful and can cause Zn toxicity to organisms. The free $Zn^{2+}$ in solution is highly toxic to plants, invertebrates, and even some vertebrate species. Acute adverse effects of high Zn intake include nausea, vomiting, loss of appetite, abdominal cramps, diarrhea, and headaches. Long-term exposure of Zn can cause chronic Zn poisoning which will cause a greatly decreased blood copper concentration, anemia, leukocyte rare disease, immunity damaged, weight loss, and other symptoms. Low concentrations of Zn are necessary for normal plant growth; however, a strong phytotoxicity of Zn and retard plant growth can appear with the presence of Zn at over 400 mg/kg. Sewage irrigating crops, especially wheat, with sewage water containing high levels of Zn, will result in uneven seedling emergence, less tiller, plant dwarf, and leaf chlorosis. The diversity of microbes in the Zn-contaminated soil also appears to decrease with increasing Zn content (Moffett et al. 2003).

## 3.2 Physiological Processes of Zinc in Plants

Zn plays a central role in healthy plant metabolism and growth processes. It is transported in xylem after being absorbed by the roots and then distributed in plant tissues. Zn is unevenly distributed in plants, and Zn concentrations in tissues could differ greatly among plant species. Zn exists in both soluble and insoluble forms in plant tissues. Carboxylic acids, amino acids, phytochelatins (PCs), nicotianamine (NA), and proteins are the main complex organic compounds with soluble Zn. Zn is also found as inorganic Zn-phosphate salts and organic Zn-phytates (White and Broadley 2011).

Concentrations of Zn in plants vary from 25 to 150 mg $kg^{-1}$ DW. The foliar Zn deficiency symptoms appear with a Zn concentration of <15 mg $kg^{-1}$ DW in leaf, while toxicity symptoms become visible when leaf Zn is over 300 mg $kg^{-1}$ DW (Broadley et al. 2007). Detrimental effects of Zn on plant can be observed if plant Zn concentration reaches 400 mg $kg^{-1}$. Nevertheless, the Zn concentration in the aerial parts can reach up to 1,000 mg $kg^{-1}$ DW for Zn-tolerant plants and even more than 10,000 mg $kg^{-1}$ DW for Zn-hyperaccumulators (Baker and Walker 1990).

As one of the most important micromineral elements, the primary physiological functions of Zn in plant biochemical activities include: (1) activating agent for vitamins; (2) participating in plant cell respiration, promoting photosynthesis via enhancing the production of chlorophyll; (3) catalyzing redox action and accelerating protein oxidation; (4) participating in the synthesis of growth hormones auxin in plants, and (5) promoting plants to thrive with high resistance for diseases and cold. Several physiological processes involved in Zn absorption, transport, and storage by plants are introduced in detail in the following sections.

### 3.2.1 Zinc Uptake from Soil by Plants

Generally, Zn can be absorbed via roots primarily as $Zn^{2+}$ and/or $Zn(OH)_2$ at high pH in the soil solution. Zn is transported either symplastically or apoplastically after being taken up through root cells (White et al. 2002; Broadley et al. 2007). The uptake of Zn by plants from soil to plant roots is proposed to be driven by the negative electrical potential in plasma membrane and mediated by the complicated metal transport systems on the plasma membrane of root cell. The availability of Zn in the soil to plant, rhizospheric process, and cell membranes transport are thought to be the important biological processes controlling Zn uptake.

The availability of Zn in soil is controlled by the factors that affect the amount of available Zn in soil solution or its sorption–desorption from/into the soil solution. These factors usually include: the total Zn content, chemical forms of Zn compounds, soil properties (organic matter content, carbonate, or phosphate content, granularity, pH), environmental conditions (temperature and humidity), concentrations of other trace elements, and relative biological activities. At low soil pH ($<6$) the bioavailability of Zn generally increases with increasing replacement of $Zn^{2+}$ by $H^+$ (Pilon-Smits 2005). A higher temperature can accelerate the biochemical activities in general, thus Zn absorption and relocation processes could be accelerated. Organic matter can either increase the Zn availability in the soil with the formation of soluble organic zinc complexes which are probably capable of absorption into plant roots or decrease its bioavailability by the formation of stable solid-state organic Zn complexes (Alloway 2008).

The rhizosphere process involves complex interactions between plant roots and rhizosphere microbes. Plant roots release a variety of organic compounds (such as organic acid, siderophores, and phenolics) that are the natural carbon resources for microbes (Bowen and Rovira 1991). At the same time, microbes in the rhizosphere stimulate root growth and enhance water and micronutrient absorption (Kapulnik 1996). There are many studies about the roles of rhizosphere microbes in protecting plant from excessive heavy metal toxicity by reducing metal absorption (Delorme et al. 2001; Whiting et al. 2001; Farinati et al. 2009). For example, under the circumstance of excessive Zn, arbuscular mycorrhiza fungi (AMF) can improve plant growth, reduce the Zn toxicity to plants, influence the absorption and translocation of Zn, and facilitate the extraction of Zn by plants from soil (Chen et al. 2003). The secretion from plants and microbes has both positive and negative effects on plant metal uptake: the secretion protects plants by reducing the absorption of metals, or promotes metal absorption by chelating metals to increase metal bioavailability.

Transport of bioavailable ions across the plasma membrane of roots is a critical step in metal uptake and accumulation. Taking $Zn^{2+}$ for example, the absorption dynamics of Zn for plants can be distinguished into two stages. At the very beginning, it is a fast and linear dynamic phase which is related to the $Zn^{2+}$ adsorption on the root cell wall. Then, there is a slower saturated adsorption stage which is related to the transport of $Zn^{2+}$ through the plasma membrane of root cells.

The Zn influx to the cytoplasm of root cells is mainly mediated by various classes of protein transporters on the plasma membrane, though some plasma membrane $Ca^{2+}$ channels are also contributing to $Zn^{2+}$ uptake (White and Broadley 2011).

Some metal transporters of the ZIP protein family are considered to be the predominant uptake systems for Zn in plants (Grotz et al. 1998; Guerinot 2000). The ZIP transporters were characterized in *Arabidopsis* (Grotz et al. 1998), soybean (Moreau et al. 2002), and rice (Ishimaru et al. 2005; Ramesh et al. 2004). The ZIP family includes a set of transport proteins, and these transport proteins have an important feature, i.e., they all can transport $Zn^{2+}$ and other metal ions from the extracellular or organelles lumen into the cytosol (Saier 1999). ZRT1 and ZRT2 are the earliest achieved ZIP family members by gene cloning. They stand for, respectively, high affinity and low affinity of the $Zn^{2+}$ absorption transporters (Zhao and Eide 1996a, b). ZRT1 and ZRT2 are responsible for the absorption of $Zn^{2+}$ across the plasma membrane, while ZRT3 in the vacuole membrane is responsible for shipping $Zn^{2+}$ from the vacuole back to the cytoplasm (Macdiarmidc et al. 2000). IRT1 and IRT2, the main $Fe^{2+}$ uptake systems in *Arabidopsis thaliana* root cells, are found to contribute significantly to the uptake of Zn and Cd by plants. An increase of IRT1 transcript and IRT1 protein levels in the root appears after the treatment of Fe limitation, which in turn leads to IRT1-dependent Cd and Zn accumulation in the roots (Palmgren et al. 2008). In addition to the ZIP transport system, yellow strip-like (YSL) proteins, which have been found mainly involved in Fe transport, are also evaluated to contribute to the uptake of Zn complexed with phytosiderophores or NA (Schaaf et al. 2005). HMA2 and HMA4 are two out of the eight gene encoding members of the type1B heavy metal–transporting subfamily of the *P*-type ATPases in *Arabidopsis thaliana* (Grotz and Guerinot 2006). They are observed to play a primary role in essential Zn homeostasis.

These transport systems, to a great extension, determine the specificity and direction of $Zn^{2+}$ transport, and the concentration of transport protein will decide the speed of $Zn^{2+}$ cross-membrane transport and the accumulation of $Zn^{2+}$ within the membrane. In Zn hyperaccumulator, *Thlaspi caerulescens*, the constitutive expression of a Zn transporter in the root cell membrane is proved to be one of the underlying mechanisms of natural hyperaccumulation. Researches on the molecular mechanism involved in Zn hyperaccumulation and hypertolerance are undergoing. Though much remains to be investigated, the existing knowledge demonstrates that the transmembrane process is one of the key factors that control essential Zn homeostasis (Clemens et al. 2002; Hussain et al. 2004).

## 3.2.2 Zinc Chelation and Compartmentation in Roots

After the transmembrane process, most of the Zn (primary $Zn^{2+}$) is chelated by several metal chelators in the cytosol or sequestrated in the vacuoles, keeping a vanishingly low concentration of free Zn ion in the cytosol. This mechanism is the main way of heavy metal detoxification in plants, contributing to metal tolerance and metal hyperaccumulating of plants. A lot of chelating materials have already been discovered for plants and metallothioneins (MTs), PCs, low-molecular chelating agents (LCs) are three typical groups of those metal chelators.

(1) MTs

MT is a kind of low molecular and cysteine-rich polypeptides found in the cytoplasm of plant cells. The hydrosulfuryl of cysteine residues can detoxify heavy metals (like Zn and Cd) by forming non-toxic or low-toxic complex (Nathalieal et al. 2001). The first MT extracted from plants is wheat EC (Early Cys) protein. It was extracted from mature embryo of wheat, and can combine with $Zn^{2+}$ (Lane et al. 1987). Since then, more than 50 kinds of MTs genes were found in different plants (Rauser 1999). MTs play an important role in chelating with heavy metals and in regulating Zn and Cd homeostasis in plant cells (Palmgren et al. 2008).

(2) PCs

PCs were first extracted from the Cd stressed *Rauvolfia serpentine* cells by Grill et al. (1985). Researches show that PCs are a kind of hydrosulfuryl chelated polypeptide, consisting of homocysteine, glutamate, and glycine, known to be required for basic tolerance of metals in all plants. Due to high contents of hydrosulfuryl, PCs have a high metal affinity, and can be chelated with many heavy metal ions. The biosynthesis of PCs can be induced by many metals, including Cd, Ni, Pb, and Zn (Kalpheck et al. 1995). After a few seconds of heavy metal processing, plant cells will induce the generation of PCs, and then the low molecular PCs in the cytoplasm will be transported to the vacuole where PCs-metals are formed and stored (Sun et al. 2005; Hangavel 2007).

(3) LCs

In addition to MTs and PCs, the third important category of metal chelating agents in plant is the LCs, including organic acids (such as oxalic acid, malate acid, citric acid, and amino acids), NA, and inorganic anions (e.g. phosphate). They play an important role in the cumulative mechanism of intracellular Zn with the ability of improving the heavy metal tolerance. LCs has been demonstrated to reduce free $Zn^{2+}$ concentration by forming chelating compounds or precipitation. Studies have shown that citric acid is related with the accumulation and resistance of $Zn^{2+}$ (Sanger et al. 1998) and malate acid is a $Zn^{2+}$ combiner in the cytoplasm (Godbold et al. 1984). NA is a ubiquitous compound in plants and has the capacity to bind Zn, Fe, and other metals (von Wiren et al. 1999; Schaaf et al. 2004). Zhao et al. showed that citric acid has a high affinity of $Zn^{2+}$, and it can form the Zn-chelating groups in *Arabidopsis halleri* (Zhao et al. 2000). In *Thlaspi caerulescens*, up to 70 % of root Zn may be associated with His (Histidine) and the remaining 30 % was associated with the cell wall (Callahan et al. 2006).

Vacuoles are assumed to be the major sites of metal sequestration in root cells (Martinoia et al. 2007). Once Zn has entered the cytoplasm of a root cell, especially in excessive concentration, it might be transported into vacuoles and sequestered there as free Zn ions or organic complex. This process is also regulated by numerous metal ion transport systems. Members of cation diffusion facilitator (CDF) family of proteins have been inferred to be the key proteins controlling this process and in contributing to Zn tolerance in plant. The CDF family members ZRC1 and COT1, found in the *Saccharomyces cerevisiae*, are related with $Zn^{2+}$ compartmentation in the vacuole by transporting $Zn^{2+}$ from the cytosol to the vacuole (Conklin et al. 2003). Maestri et al. (2010) revealed that the MTP1 genes from *Thlaspi caerulescens, Arabidopsis halleri,* and *Thlaspi goesingense* are involved in increasing Zn sequestration by promoting the influx of Zn in the vacuole. Recent evidences suggest that expression of MTP1 might also respond to Zn deficiency, leading to increasing Zn uptake and accumulation (Gustin et al. 2009). Another protein MTP3 in *Arabidopsis* was found to play an essential role in $Zn^{2+}$ tolerance and compartmentation (Arrivault et al. 2006). Other members of CDFs family, e.g., the $Mg^{2+}/H +$ antiporter AtMHX and the orthologs of the *Arabidopsis thaliana* Zn-induced facilitator 1 (AtZIF1) protein, can transport $Zn^{2+}$ and $Zn^{2+}$-complexes separately into the vacuole (Maestri et al. 2010).

### 3.2.3 Translocation of Zinc from Root-to-Shoot

Water and mineral nutrients are primarily transported through the xylem in plant. Within the xylem, Zn is present and transported predominantly as $Zn^{2+}$ and a complex with organic acids or NA. It was suggested that the xylem loading be the committed step in root export of metal ions to the shoot, partially controlling the concentration and distribution of Zn in plants. A series of transmembrane activities exist in these procedures that are regulated by variety of metal transporters intracellular or on the plasma membrane.

Phenotypic analysis of *Arabidopsis thaliana* mutants which carry the disruptions of genes HMA2 and HMA4 that code for two HMA (HMA: Heavy Metal Transporting ATPase) Zn pumps in the root parenchyma provides the strong evidence to confirm that xylem loading is the key step for Zn translocation from root to shoot (White and Broadley 2011). These HMAs are supposed to transport Zn across the plasma membrane of root vascular cells into the xylem prepared for transport to the shoot (Palmgren et al. 2008). The results revealed that the hma2 and hma4 genes double mutant greatly increases Zn contents in roots and decreases Zn contents in shoots, indicating the fact that relative gene coded Zn pumps are required for Zn translocation upwards (Hussain et al. 2004). Increasing the expression of Zn pumps coding genes like HMA2 and HMA4 in the *Arabidopsis thaliana* could be a positive attempt to enhance the rate of the root-to-shoot Zn translocation in Zn phytoremediation plants (i.e., the translocation factor). YSL proteins are also contributing to load Zn-NA complexes into the xylem (Curie et al. 2009). The Zn sequestered in the root vacuole is thought to be released through another class of metal transporter, members of the natural resistance-associated macrophage proteins (NRAMPs) family, including orthologs of the AtNRAMP3 and AtNRAMP4 transporters of *Arabidopsis thaliana* (Roosens et al. 2008; Verbruggen et al. 2009).

### 3.2.4 Zinc Distribution and Storage in Aerial Parts of Plant

Because of an effective metal excretion system lacking in plants, excessive Zn is transported to and compartmented in certain inactive metabolic organs or sub-cellular areas to avoid an excessive cytoplasmic Zn which can result in direct toxicities to the plant. Compartmentation of metals such as Zn, Cd, and Ni in the aerial parts has been regarded as one of the most probable mechanisms of metal detoxification in hyperaccumulating plants (Broadley et al. 2007).

Micronutrients like Zn have to be unloaded from the xylem and be actively taken up by living cells surrounding the xylem in order to enter the mesophyll in leaf cells and other organs. From these cells, micronutrients move from cell to cell until they reach their final destination in plants for their normal physiological functions or, if in excessive levels, are stored in specific locations like cell wall or vacuoles. Xylem-unloading processes are thought to be the first step in controlled distribution and detoxification of metals in shoots, as well as in a possible redistribution of metals via the phloem afterward (Schmidke and Stephan 1995). The distribution of metal ions within the leaf after unloading them from the xylem is found to be via the apoplastic or symplastic passage (Karley et al. 2000). The uptake of Zn by specific cell types within shoots is also facilitated by certain metal transport proteins. Members of the ZIP family are thought to mediate $Zn^{2+}$ influx to leaf cells and also involve in the Zn loading into the phloem while YSL proteins are implicated in the same physiological process of Zn-complex in shoots (White and Broadley 2011).

Generally, concentrations of Zn in plant cells are within the specific physiological ranges, while excessive Zn can be sequestered in the aerial part of Zn hyperaccumulating plants. Intriguingly, the forms and distribution pattern vary between plant species. It was reported that in *Athyrium yokoscense* 70–90 % of Zn was stored in the cell wall in the form of ionic compounds or directly combined with the cell wall materials (Nishizono et al. 1987). However, in *Arabidopsis halleri* and *Thlaspi caerulescens,* the major Zn storage compartment is mesophyll tissues and vacuoles (Kupper et al. 1999; 2000).

Metal chelation in leaves shares the similar chelating mechanisms in the roots. A large number of metal chelators like organic acid, glutathione, PCs, NA, and proteins are produced or transported into the specific sites where excessive Zn exists, forming metal chelating complexes to keep a relatively steady Zn homeostasis in plant. The detailed mechanisms have been discussed in Sect. 2.2.

## 3.3 Phytoremediation of Zinc-Contaminated Soils

### 3.3.1 Zinc Contamination in Soil

With thousands of years of mining and processing, anthropogenic emission of Zn has become a primary cause for environmental Zn contamination. Zn is the fourth most common metal in use (after iron, aluminum, and copper) with an annual production of about 12 million tons (Tolcin 2011). About 70 % of the world's Zn originates from mining, while the remaining 30 % is from recycling. Globally, the major Zn mining countries are China, Australia, and Peru. China contributes 29 % to the global Zn production (Tolcin 2011). Zn is released into the environment through fossil fuel combustion, mine waste, phosphate fertilizers, limestone, manure, sewage sludge, and particles from galvanized surfaces. Soil Zn concentrations of over 1,000 mg kg$^{-1}$ DW have been found in certain agricultural fields near industrial sites, compared with a background concentration of 100 mg kg$^{-1}$ DW in agricultural soils (Audet and Charest 2006). Excessive Zn of 500 mg kg$^{-1}$ in soil interferes with the ability of plants to absorb other essential elements, such as iron and manganese. Zn levels of 2,000–180,000 mg kg$^{-1}$ (or 18 %) have been recorded in some Zn-contaminated soil samples (Emsley 2001).

### 3.3.2 Phytoremediation

Soil heavy metal contamination can be remediated by different chemical, physical, and biological techniques. The application of physical and chemical remediation technologies at contaminated sites likely cause adverse impacts on the ecosystem and generally the cost is relatively high for remediating a large polluted area. Among the numerous currently available remediation techniques, phytoremediation receives more attention as a cost-competitive, environmental-friendly, esthetically pleasing approach for site remediation.

Phytoremediation is using plants to take up, accumulate, store, degrade, or render organic or inorganic contaminants in contaminated soil and water by taking advantage of the natural abilities of plants. In general, phytoextraction, phytostabilization, phytodegradation, phytostimulation, and phytovolatilization are the five main subsets of phytoremediation that have been identified (Fig. 3.1) among which phytoextraction and phytostabilization are the most commonly applied processes for metal remediation. When applied in contaminated sites, one restriction to be considered is that phytoextraction is potentially feasible only in soils with low or moderate levels of contamination. However, for the heavily contaminated sites, phytostabilization with tolerant plants may be a more suitable strategy by stabilizing the contaminated sites and reducing the risk of erosion and leaching of pollutants (McGrath and Tunney 2010). Continuous or natural phytoextraction and chemically enhanced phytoextraction are two approaches that

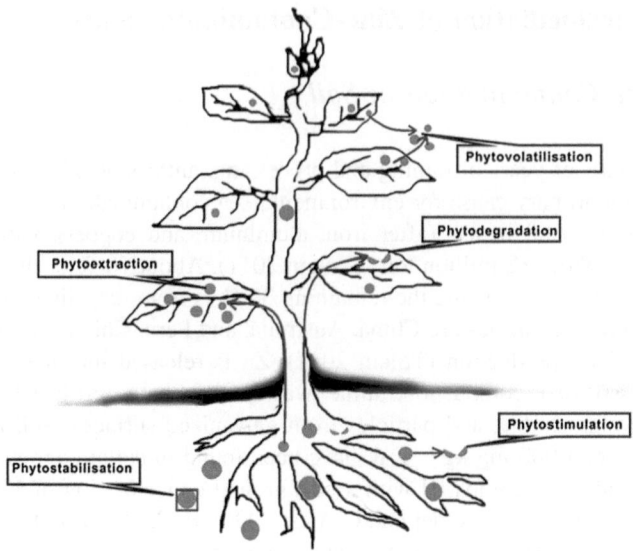

**Fig. 3.1** Schematic diagram of main subsets of phytoremediation

have been proposed for phytoextraction of heavy metals (Lombi et al. 2001; Ghosh and Singh 2005)

Phytostabilization is a technique that can reduce the mobility and bioavailability of metal pollutants in the soil, aiming at decreasing the risks of pollutants to human health and the environment. Soil amendments, such as phosphate fertilizers, organic matter, Fe- and Mn-oxyhydroxides, and inorganic clay minerals can be applied to further enhance the reduction of metal bioavailability, preventing plants from absorbing or transporting in the surrounding environment (Berti and Cunningham 2000). Plants having extensive and abundant roots primarily accumulate pollutants in roots which are good candidate species for phytostabilization (Mendez et al. 2008).

Compared with phytostabilization, phytoextraction exploits the ability of plants to translocate a great fraction of metals taken up for harvesting biomass. Favorable plant properties for phytoextraction are generally fast growing, with high biomass production, an extended root system, and high translocation factor (TF, shoot-to-root metal concentration ratio), accumulation in harvestable tissues, and also easy agricultural management (Vamerali et al. 2010). Since it is more reliable to remove metals directly out of the contaminated sites than other physiochemical technologies, much interest is devoted to this technology and its improvement, while a better understanding of the physiological and molecular mechanisms in hyperaccumulators has inspired further improvement of the phytoremediation technology.

There are both advantages and disadvantages to the use of phytoremediation. Phytoremediation is less expensive than the traditional methods that clear the

**Table 3.1** General physiological features observed in Zn hyperaccumulators

| Different parts in plants | General features of Zn hyperaccumulators |
| --- | --- |
| In roots | Zn uptake rates are increased, generally reflected as an increased maxima velocity of absorption; |
| | Zn sequestration in the roots is decreased which may be induced by an enhanced root-to-shoot Zn transport; |
| | The rate of loading of Zn from the root into the xylem for root-to-shoot transport is strongly enhanced; |
| In shoots | Zn sequestration in the shoots, mostly inside the vacuoles of leaf cells, is highly effective; |
| | Cellular Zn uptake rates are enhanced, with a highly effective system for cell-to-cell Zn distribution. |

contaminated sites by pumping, washing, or digging soil out of the contaminated site. This property makes phytoremediation even cheaper than traditional remediation methods. Since most plants used for phytoremediation are originally discovered on metal contaminated and uninhabitable sites, the wildlife there is able to flourish after being treated by those pioneers. But on the other hand, phytoremediation is restricted by the root depth of plants being used and can function only with low-to-moderate levels of contamination in the field. It may take many years to remediate the contaminated sites, so it is a time-consuming remediation method. Besides, low biomass and management of those applying plants are also the bottlenecks that restrict the implementation of this technology.

### 3.3.3  Zinc Hyperaccumulation

The first Zn hyperaccumulator species was reported in 1865, when Baumann et al. found that the Zn oxide contents reached up to 17 % in the ash of stem and leaf of *Thlaspi calaminare* (Sachs 1865). Since then more Zn hyperaccumulators have been identified. Baker et al. (1989) defined the plants that accumulate Zn more than 10,000 mg kg$^{-1}$ DW (or 1 %, w/w) in the aerial parts as Zn hyperaccumulators. However, Broadley et al. (2007) recently suggested that 3,000 $\mu$g g$^{-1}$ Zn concentrations (DW) in the aerial part might be more realistic. So far, among the reported metal hyperaccumulators, there are about 15–20 species for Zn, most of them in Brassicaceae (White and Broadley 2011). In addition to *Thlaspi calaminare*, *Arabidopsis halleri* is another most frequently studied Zn hyperaccumulator, which advances our understanding of Zn hyperaccumulation and hypertolerance at cellular and molecular levels (Kramer 2010). Yang et al. (2002) identified *Sedum affredii Hance* as the potential Zn hyperaccumulator in China, with an average aboveground Zn concentration of 4,515 mg kg$^{-1}$ DW in the field and the highest shoot Zn of 19,674 mg kg$^{-1}$ DW in nutrient solution. The physiological processes of Zn uptake, transport to the xylem, and tolerance in shoot tissues are maximized in Zn hyperaccumulators. Certain features of Zn hyperaccumulators are summarized in Table 3.1.

Molecular mechanisms of Zn tolerance and hyperaccumulation are closely related to a set of constitutively highly expressed genes encoding metal transporters in the plasma membrane and enzyme-catalyzing compounds synthesis in physiological processes that facilitate Zn hyperaccumulating in those plants. So far, these Zn transporters have been found in several protein families, including ZIPs, YSLs, HMAs, ZIF1, MTPs, NRAMP (Yang et al. 2005; White and Broadley 2011). Compounds synthesized under certain enzymes catalysis (such as phytosiderophores and NA) are found to regulate Zn homeostasis by controlling the cell-to-cell mobility of Zn, excessive Zn sequestration in vacuoles, or other specific parts of plant cells. Taking one of the Zn hyperaccumulators, for example, upon comparison of *Thlaspi caerulescens* and *Arabidopsis thaliana*, ZIP4, ZIP10, and especially IRT3 (all from ZIP family proteins) were found to be much higher expressed in *Thlaspi caerulescens* roots than in *Arabidopsis thaliana* roots, even at different Zn exposures (van de Mortel et al. 2006). More molecular mechanisms and examples that are related to other metal transporters and enzymes regulation have been included in the part of "Physiological processes of Zn in plants".

## 3.3.4 Zinc Phytoremediation Strategies

Soil Zn contamination is a typical instance of metal pollution that can be cleaned up by phytoremediation. Many studies have been carried out both in the laboratory and in the field to develop Zn phytoremediation technology, but no successful full-scale application has been reported yet.

### 3.3.4.1 Natural Zinc Phytoremediation

Phytoremediation was first intended to utilize the natural properties of Zn-tolerant or hyperaccumulating plants to remove excessive Zn in the soil. Although promising, using this approach to remediate Zn contaminated sites is faced with many difficulties in the practical application. Lacking information on the agricultural management, slow growth rate, poor biomass of whole plants, and consuming a fairly long time and so on are only some of the barriers that block this application. Robinson et al. (1998) evaluated Zn uptake by *Thlaspi caerulescens* in pot trials and in wild populations at a mine waste site in France. It was estimated that the plant could remove 60 kg of Zn per ha per cropping, which was considered to be insufficient for Zn remediation. Brown et al. evaluated cadmium and Zn uptake by *Thlaspi caerulescens*, silene, and lettuce in 2-year-long field studies. The soils were contaminated by sewage sludge. A total of 18 growing-seasons will be needed to remediate a soil containing 400 mg Zn $kg^{-1}$ (Brown et al. 1995).

Though most high yielding plant species have a relatively low tolerance for Zn, certain high biomass, and fast growing plant species may be the potential candidate species. Vamerali et al. (2010) reported that *Brassica*, *Zea mays*, and *Phaseolus vulgaris L.* could accumulate more than 1,000 mg Zn/kg DW. Sunflowers and maize have strong ability to take up Zn and other metals from the soils (Fellet et al. 2007; Tassi et al. 2008). Despite the superiority of certain field crops in high biomass compared to hyperaccumulator plant species, phytoremediation of Zn-polluted soils is still limited by the plant species under field conditions.

### 3.3.4.2 Improving Zinc Phytoremediation Efficiency

Plants used for phytoremediation are expected to have both high metal accumulation in shoots and high shoot biomass production (Vamerali et al. 2010). However, this ideal plant species has not been discovered. For this reason, promising physical, chemical, agricultural, and biotechnological approaches for enhancing the potential for Zn phytoremediation are explored either to improve the growth rate or biomass of phytoremediation plants. Phytohormones can be applied to promote root development and growth and further increase the whole plant growth or biomass production.

Chelating agents for enhancing phytoremediation (or phytoextraction) have been investigated to increase the heavy metal accumulation in plants having high biomass production and metal tolerance. Previous studies reported dramatic increases in plant Zn accumulation from soil in the presence of added synthetic chelates (Blaylock et al. 1997). The following order of extraction efficiency to Zn was achieved using biodegradable chelating agents to extract heavy metals from soil: NTA > EDDS > EDTA > MGDA > IDSA (Tandy et al. 2004). Unfortunately, without the appropriate management, using chelating agents to enhance phytoremediation could result in certain side effects, such as heavy metal leaching, reduced microbial diversity, and accumulation of refractory organic chemical chelating agents in the environment (Römkens et al. 2002).

Genetic engineering techniques that focus on improving growth and biomass production of known Zn hyperaccumulators are undergoing, though no successful case under field condition has been reported yet. Several targets are suggested for genetic engineering to improve Zn tolerance/accumulation in normal plant species, including overexpression of natural metal chelators (MTs, PCs, and LCs); regulating the metal transport systems on plasma membrane; alteration of the Zn metabolic pathways. Desbrosses-Fonrouge et al. (2005) identified the genes responsible for metal hyperaccumulating properties in model Zn/Cd hyperaccumulator *Thlaspi caerulescens*, which, if well characterized and properly expressed, could transform the high biomass producing species into metal tolerant and Zn accumulation. In an effort to correct for small sizes of hyperaccumulator plant, somatic hybrids have been generated between *Thlaspi caerulescens* and *Brassica napus*. The high biomass hybrid selected for Zn tolerance is capable of accumulating Zn to the level that would have been toxic to *B. napus* (Brewer et al. 1999).

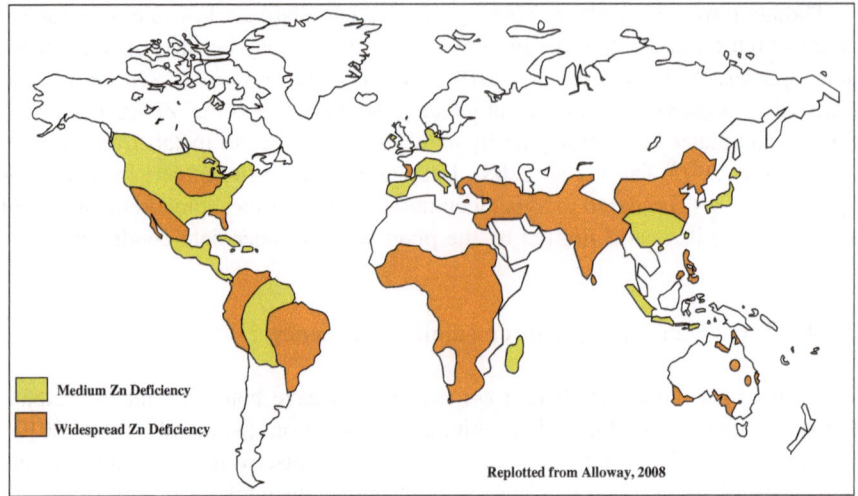

**Fig. 3.2** Global distribution of Zn-deficiency affected regions (Replotted from Alloway 2008)

More studies will be needed to better understand the molecular and genetic mechanisms in hyperaccumulators of Zn and other metals, for setting the stage for the feasible and effective application of phytoremediation.

## 3.4 Zinc Biofortification for Human Nutrition

### 3.4.1 Zinc Deficiency in Human Body

In the early 1960s, Zn deficiency in the human body was first speculated with considerable supportive evidence (Prasad et al. 1963). Nowadays, nearly two billion people in the developing world are suffering from Zn deficiency. The global distribution of Zn-deficiency affected regions are shown in Fig. 3.2. Soil Zn deficiency is among the major global micronutrient deficiencies and has recently received more and more attention.

Zn deficiency is responsible for many health problems, including impairments of physical and mental growth, immune system, high risk of infections, DNA damage, and cancer development (Hotz and Brown 2004; Gibson 2006; Prasad 2007). The public health implications of Zn deficiency in the developing world are being for pursued for decades. It is now well established that Zn deficiency is responsible for diarrhea and pneumonia in children (Gibson et al. 2008; Walker and Black 2009). Retarded growth and dwarfishness are widely studied and considered to be the indicators of human Zn malnutrition especially in infants and young children. Besides, pregnant women are another group susceptible for Zn deficiency and a survey among 285 pregnant women in Haryana showed that 65 %

**Table 3.2** RDAs for Zn

| Age | Male (mg) | Female (mg) | Pregnancy (mg) | Lactation (mg) |
|---|---|---|---|---|
| 0–6 months | 2[a] | 2[a] | | |
| 7–12 months | 3 | 3 | | |
| 1–3 years | 3 | 3 | | |
| 4–8 years | 5 | 5 | | |
| 9–13 years | 8 | 8 | | |
| 14–18 years | 11 | 9 | 12 | 13 |
| 19 + years | 11 | 8 | 11 | 12 |

[a] Adequate intake (AI)

of them suffer from Zn deficiency (Prasad 2010). Improving Zn malnutrition for the worldwide population, especially in developing countries has become an urgent task which calls for close international communication and cooperation between governments and research institutes.

Past efforts on agriculture production have primarily been focused on increasing crop yields; however, the accompanying decrease of mineral concentrations in grains was found as a new problem threatening the development of crop yields and even the food security. The ultimate goal of modern agriculture has been modified to produce nutritious foods sufficiently and sustainably (Zhao and McGrath 2009). The contents of nutrients in the edible parts of staple food crops, e.g. maize, rice, wheat, barley, contribute to the main mineral intake of people in the developing countries. Therefore, increasing concentrations of mineral elements, like Zn, Fe, and Se, in staple food crops is the most effective approach for public health to control malnutrition in Zn deficiency areas.

Recommended Dietary Allowances (RDAs) for Zn developed by the Food and Nutrition Board (FNB) at the Institute of Medicine of the US National Academies are displayed in Table 3.2 (Institute of Medicine, Food, and Nutrition Board 2001).

Although animal products, such as meat, fish, and poultry, contain more Zn than cereals, plant foods with low concentrations of Zn almost occupy the main foods in low developed countries. Zn concentrations in even the most favorable plant foods are inadequate to meet the requirements (Gibson et al. 1998). This problem is compounded by the limiting content of Zn and low Zn bioavailability in vegetarian diets.

## 3.4.2 Zinc Biofortification Strategies

To overcome Zn deficiency in humans, there are two main approaches: (i) changing dietary composition through dietary Zn supplements and (ii) biofortification on staple foods through increasing the Zn content of food grains by plant breeding. A wide public awareness and sustained funding from the government are required by the first approach, which makes it not so easy. Biofortification has no such difficulties and meantime dominates on the feasibility to be applied in both urban and rural areas. For example, for the millions of Zn-deficient people in South Asia, the major daily consumed staple crops, rice, and wheat, are chosen as the candidate for Zn biofortification to improve local malnutrition.

Biofortification is defined as a technology to improve the micronutrition contents in staple crops using traditional breeding practices and modern biotechnology. One way is genetic biofortification, including conventional breeding and genetic modification (GM), and the other is agronomic biofortification embodied as the application to micronutrients in fertilizers.

### 3.4.2.1 Breeding Strategies for Zinc Biofortification

With the abundant natural genetic variations and centuries of conventional breeding experiences, plant breeding strategy is widely accepted as a cost-effective and easily affordable solution among the stable food crops biofortification approaches. It is possible that breeding can increase Zn-tolerance in root and leaf crops and increase Zn mobility in the phloem of fruit, seed, and tuber crops. Currently, numbers of breeding programs are ongoing aimed at developing new cereal genotypes with high genetic ability to absorb Zn and also other micronutrients from soil and finally accumulate Zn in grain at desired levels. Plenty of organizations and research institutes are devoted to this challenging task, among which the HarvestPlus-Biofortification Challenge Program is one of the leading programs, aiming at improving stable food crops with Zn, Fe, and vitamin A, by using plant breeding strategy (Pfeiffer and McClafferty 2007).

Genetic variations in grain are essential for the development of new genotypes with high Zn concentrations. However, since cultivated crops contain narrow genetic variation for Zn accumulation, species with promising genetic resource for higher Zn concentration are needed. In a series of genetic variation collections of wild emmer wheat, quite a few wide wheat varieties are found with not only high concentrations of Zn in seeds but also with high tolerance to drought and Zn deficiency in soil (Peleg et al. 2008). A large genetic variation also exists in grain Zn concentration in different germplasms of other crops such as rice and maize, and related researches are undergoing in certain breeding programs (Graham et al. 1999).

In addition to traditional breeding, genetic breeding through GM also attracts the attention of researchers. The identification of certain genes that control physiological activities such as Zn uptake, translocation, distribution, and sequestration in plant (especially in some Zn-tolerant plants and Zn hyperaccumulators), together with the confirmation of numerous enzymes involved in the Zn homeostasis in plant, contribute to better understanding of the mechanism for Zn tolerance and accumulation in plant and also provide the theoretical basis for genetic breeding of grains with higher Zn concentration. A research about the transcription factors that regulate the adaptation to Zn deficiency in *Arabidopsis thaliana* speculates that the overexpression of bZIP19 and bZIP23 transcription factors could be used to increase Zn accumulation in edible portions of crops by inducing constitutive expression of a suite of Zn-deficiency responses (Assuncao et al. 2010). In *Arabidopsis thaliana*, when reducing the expression of AtHMA2, which is speculated to catalyze $Zn^{2+}$ efflux across the membranes of root cells (Eren and Arguello 2004) or overexpressing the gene encoding AtHMA4, which is thought to load Zn into the xylem (Verret et al. 2004), and an increased leaf Zn concentration was observed. In another example, after the overexpression of HvNAS1 in tobacco, Zn concentrations in leaf and seed increased from 16 to 39 mg $kg^{-1}$ DW and from 20 to 35 mg $kg^{-1}$ DW, respectively (Takahashi et al. 2003). Several transgenic plants that have greater Zn concentrations in their edible tissues than conventional varieties have been created. A variety of cassava with roots Zn 40 mg $kg^{-1}$ DW (Sayre et al. 2011), the brown rice with 56–95 mg Zn $kg^{-1}$ DW (Vasconcelos et al. 2003; Johnson et al. 2011), and barley grain with 85 mg Zn $kg^{-1}$ DW (Ramesh et al. 2004) have been reported.

However, in the implement of crops breeding, various environmental conditions can affect the effects of Zn biofortification during the long-term process over years. The adverse chemical and physical properties of cultivated soils reduce chemical solubility and availability of Zn in soils, resulting in inadequate amounts of Zn absorption from soils. Among the chemical factors, high soil pH is among the most critical factors reducing solubility and root absorption of Zn. Changing soil pH from 6 to 7 results in about a 30-fold decrease in soil Zn solubility and further significantly decreases plant Zn concentrations. Similar impairments in root absorption of Zn also take place in soils with low levels of soil moisture and organic matter. In Turkey, low annual rainfall (<300 mm), relatively high soil pH (7.5–8.1), and low soil organic matter (averaged 1.5 %) are responsible for the severe Zn deficiency in Central Anatolia (Cakmak et al. 1999). In many other cultivated soils around the world, such as in China, India, Iran, Pakistan, and Australia, similar Zn deficiency in soils has been also reported. Under such adverse environmental conditions, the newly cultivated plant with ability to accumulate high Zn in edible parts may not achieve the desired effect.

### 3.4.2.2 Agronomic Biofortification

Agronomic biofortification is generally known as the application of Zn fertilizers to soil and/or foliar to increase grain Zn concentrations. It is considered to be a flexible approach that can be used for all crop species and cultivars, and meantime an important complementation to the ongoing breeding programs of cereals with high Zn in grain. Compared to genetic biofortification, it is considered as a short-term solution without years of crossing and backcrossing activities. To gain high Zn in the grain by application of Zn fertilizers, two general conditions are required: first, keeping sufficient amount of available Zn in soil solution; second, maintaining adequate Zn transport to the seeds during the reproductive growth stage. The most widespread inorganic Zn fertilizer is zinc sulfate, along with zinc oxide

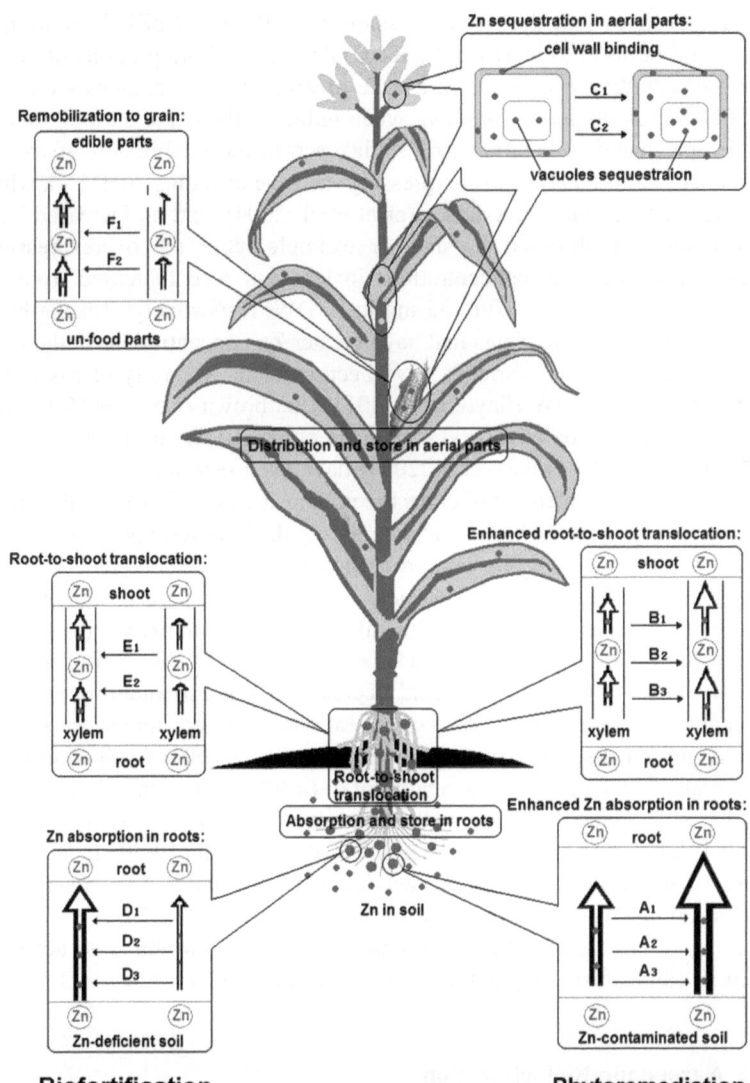

**Fig. 3.3** Key processes and potential improvements in phytoremediation of Zn-contaminated soil and Zn-biofortification. $A_1$:Zn hyperaccumulators inherently absorb excessive Zn from soil to root; $A_2$: Increasing Zn bioavailability (1, decreasing soil pH; 2, enhancing microbial activities; 3, using synthetic Zn chelators); $A_3$: Highly expression Zn transporter genes on root cell membrane; $B_1$: Reduced Zn sequestration in roots; $B_2$: Enhanced Zn pumping into xylem (mediated by HMAs); $B_3$: Strong Zn demand signal in the shoots; $C_1$: Increased synthesis of Zn chelators in leaf cells; $C_2$: Excessive Zn sequestration in specific positions (vacuoles, cell wall); $D_1$: Application of Zn fertilizers; $D_2$: Increasing Zn bioavailability (similar to $A_2$); $D_3$: Traditional or genetic breeding to produce crops species with high Zn uptake ability; $E_1$: Identify traits responsible for efficient Zn translocation to shoots; $E_2$: Genetic transfer of potential governing factors of Zn xylem loading in Zn hyperaccumulators; $F_1$: Remobilization of Zn in non-food parts; $F_2$: loading more Zn into edible parts (like grains, fruits)

and synthetic Zn-chelates (White and Broadley 2009). Although the agronomic effectiveness of Zn fertilizers is higher with Zn-EDTA than the inorganic Zn fertilizer, its high cost and potential environmental adverse effects limit the use of Zn-EDTA in cereal farming. Safe and accurate application systems are required to process the fertilizers. When Zn fertilizers are applied in the soil, Zn phyto-availability and acquisition by roots should be considered for better Zn uptake. In addition to delivering phytoavailable Zn-fertilizer to the soil or foliage, certain agronomic strategies like reducing soil pH, adopting appropriate crop rotations, or introducing beneficial soil microorganisms contribute to high Zn phytoavailability (Rengel 1999; He and Nara 2007; White and Broadley 2009). When Zn-fertilizers are applied to foliage, it is particularly important to use soluble Zn compounds, making sure they can enter the leaf apoplast, and can be taken up by plant cells without having buildup on the foliar surface (Haslett et al. 2001; Cakmak 2008; Brown 2009).

There are a few convincing evidences on the role of Zn fertilizer application in improving grain Zn concentration. In Turkey, Zn biofortified wheat has been produced in field trials. Applying Zn fertilizers to wheat grown in fields in Central Anatolia not only improved productivity, but also increased grain Zn concentration (Yilmaz et al. 1997). Depending on the application method, Zn fertilizers can increase grain Zn concentrations up to 3- or 4-fold. The most effective method for increasing Zn in grain was to combine both soil and foliar application. When a high concentration of grain Zn is aimed in addition to a high grain yield, combined soil, and foliar application is recommended. Alternatively, using seeds with high Zn concentrations at sowing together with foliar application of Zn is also an effective way to improve both grain yield and grain Zn concentrations. New research programs are needed to develop or improve Zn application methods in terms of form, dose, and application time of Zn fertilizers.

## 3.5 Summary

Phytoremediation of Zn is an interdisciplinary technology that can benefit from many different approaches. The processes affecting metal availability, metal uptake, translocation, and chelation need further study. Since the inevitable involvement of engineering, biology, agronomic management even the use of chemical substances in the process of Zn phytoremediation, effective evaluation, and prevention strategies must be developed to avoid the possible negative impacts.

To increase Zn concentrations in edible crops by biofortification strategy, agronomic and genetic strategies should be integrated in the future. At the same time, a further identification of the mechanisms effecting general homeostatic regulation of tissue Zn concentrations within the plant and strategy of effective sequestration of Zn in non-vital compartments is required for the development of more feasible biofortification strategies. For this, a variety of lessons could be learnt from the studies of Zn phytoremediation especially those about Zn-tolerance and Zn-hyperaccumulation plants. The combination of breeding and fertilizer strategies is an excellent complementary approach to alleviate Zn-deficiency-related problems in human nutrition.

Both environmental Zn contamination and Zn malnutrition in humans are globally challenging problems that call for the concerted efforts of researchers in multiple fields, including plant biology, plant breeding, biotechnology, nutrition, and environmental sciences. Phytoremediation and biofortification of Zn as the promising representative solution of each problem should have a mutual reference on mechanism and application. The simplified relation and distinction between phytoremediation of Zn contaiminated soil and biofortification of Zn for human nutrition are shown in Fig. 3.3.

Researches on Zn phytoextraction mainly take typical plants with the ability of Zn hyperaccumulating or Zn tolerance as the studying objects, but never stop the exploration of phytoremediating plants from general plants especially cereal plants with high biomass. For biofortification, cultivating cereal crops with satisfactory bioconcentration of Zn in the edible parts should take advantage of the existing positive research results of phytoremediation on the aspects of enhancing trace element bioavailability in the rhizosphere, translocation from roots to shoots, and further toward grain tissues. In both cases, significant progress can only be made through a better understanding of the underlying mechanisms of Zn acquisition, transport, and homeostasis in plants.

# References

Alloway BJ (2008) Zinc in soils and crop nutrition. IZA Publications. International Zinc Association, Brussels, pp 1–116

Arrivault S, Senger T, Krämer U (2006) The Arabidopsis metal tolerance protein AtMTP 3 maintains metal homeostasis by mediating Zn exclusion from the root under Fe deficiency and zinc over supply. Plant J 46:861–879

Assuncao AGL, Herrero E, Lin YF et al (2010) *Arabidopsis thaliana* transcription factors bZIP19 and bZIP23 regulate the adaptation to zinc deficiency. Proc Natl Acad Sci U S A 107:10296–10301

Audet P, Charest C (2006) Effects of AM colonizationon "wild tobacco" plants grown in zinc-contaminated soil. Mycorrhiza 16:277–283

Auld DS (2001) Zinc coordination sphere in biochemical zinc sites. Biometals 14:271–313

Baker AJM, Brooks RR (1989) Terrestrial higher plants which hyperaccumulate metallic elements—a review of their distribution, ecology and phytochemistry. Biorecovery 1(2):81–126

Baker AJM, Walker PL (1990) Ecophysiology of metal uptake by tolerant plants. In: Shaw AJ (ed) Heavy metal tolerance in plants. CRC Press, Boca Raton, pp 155–178

Berti WR, Cunningham SD (2000) Phytostabilization of metals. In: Raskin I, Ensley BD (eds) Phytoremediation of toxic metals: using plants to clean up the environment. Wiley, New York, pp 71–88

Blaylock M, Salt DE, Dushenkov S et al (1997) Enhanced accumulation of Pb in Indian mustard by soil-applied chelating agents. Environ Sci Technol 31:860–865

Bowen GC, Rovira AD (1991) The rhizosphere—the hidden half of the hidden half. In: Waisel Y, Eshel A, Kaffkafi U (eds) Plant roots—the hidden half. Marcel Dekker, New York, pp 641–669

Brady JE, Humiston GE, Heikkinen H (1983) General chemistry: principles and structure, 3rd ed. Wiley, p 671. ISBN 047186739X

Brewer EP, Saunders JA, Angle JS et al (1999) Somatic hybridization between the zinc accumulator Thlaspi caerulescens and Brassica napus. Theor Appl Genet 99:761–771

Broadley MR, White PJ, Hammond JP et al (2007) Zinc in plants. New philol 173:677–702

Brown P (2009) Development of a model system for testing foliar fertilizers, adjuvants and growth stimulants. In: Proceedings of the California department of food and agriculture fertilizer research and education program conference 2008. Visalia, CA, pp 17–23

Brown SL, Chaney RL, Angle JS et al (1995) Zinc and cadmiumuptake by hyperaccumulator Thlaspi caerulescens grown in nutrient solution. Soil Sci Soc Am J 59:125–133

Cakmak I (2000) Role of zinc in protecting plant cells from reactive oxygen species. New Phyl 146:185–205

Cakmak I (2008) Enrichment of cereal grains with zinc: agronomic or genetic biofortification? Plant Soil 302(1–2):1–17

Cakmak I, Kalayci M, Ekiz H et al (1999) Zinc deficiency as an actual problem in plant and human nutrition in Turkey: a NATO-science for stability project. Field Crops Res 60:175–188

Callahan DL, Baker AJM, Kolev SP et al (2006) Metal ion ligands in hyperaccumulating plants. J Biol Inorg Chem 11:2–12

Chen BD, Li XL, Tao HQ et al (2003) The role of arbuscular mycorrhizal in zinc uptake by red clover growing in a calcareous soil spiked with various quantities of zinc. Chemosphere 50:839–846

Clemens S, Palmgren MG, Krämer U (2002) A long way ahead: understanding and engineering plant metal accumulation. Trends Plant Sci 7(7):309–315

Conklin DS, Mcmaster JA, Culbertson MR et al (2003) COT 1, a gene involved in cobalt accumulation in Saccharomyces cerevisiae. Mol Cell Biol 12:3678–3688

Curie C, Cassin G, Couch D et al (2009) Metal movement within the plant: contribution of nicotianamine and yellow stripe1-like transporters. Ann Bot 103:1–11

Delorme TA, Gagliardi JV, Angel FS, Chaney RL (2001) Influence of the zinc hyperaccumulator Thlaspi caerulescens J & C Presl and the nonmetal accumlator Trifolium Pratense L. On soil microbial populations. Can J Microbiol 47:773–776

Desbrosses-Fonrouge AG, Voigt K, Schröder A et al (2005) Arabidopsis thaliana MTP 1 is a Zn transporter in the vacuolar membrane which mediates Zn detoxification and drives leaf Zn accumulation. Fed Eur Biochem Soc 579(19):4165–4174

Emsley J (2001) "Zinc". Nature's building blocks: an A–Z guide to the elements. Oxford University Press, Oxford, pp 499–505. ISBN 0-19-850340-7

Eren E, Arguello JM (2004) Arabidopsis HMA2, a divalent heavy metal-transporting PIB-typ ATPase, is involved in cytoplasmi $Zn^{2+}$ homeostasis. Plant Physio 136:3712–3723

Farinati S, DalCorso G, Bona E et al (2009) Proteomic analysis of Arabidopsis halleri shoots in response to the heavy metals cadmium and zinc and rhizosphere microorganisms. Proteomics 9:4837–4850

Fellet G, Marchiol L, Perosa D, Zerbi G (2007) The application of phytoremediation technology in a soil contaminated by pyrite cinders. Ecol Eng 31:207–214

Ghosh M, Singh SP (2005) A review on phytoremediation of heavy metals and utilization of its byproducts. Appl Ecol Environ Res 3(1):1–18

Gibson RS (2006) Zinc: the missing link in combating micronutrient malnutrition in developing countries. Proc Nutr Soc 65:51–60

Gibson RS, Ferguson EL, Lehrfeld J (1998) Complementary foods for infant feeding in developing countries: their nutrient adequacy and improvement. Eur J Clin Nutr 52:764–770

Gibson RS, Hess SY, Hotz C et al (2008) Indication of zinc status at the population level, a review of the evidence. Brit J Nutr 99:14–23

Godbold DL, Horst WJ, Collins JC et al (1984) Accumulation of zinc and organic-acid in roots of zinc tolerant and non-tolerant ecotypes of deschampsia-caespitosa. J Plant Physiol 116(1):59–69

Graham R, Senadhira D, Bebe S et al (1999) Breeding for micronutrient density in edible portions of staple food crops: conventional approaches. Field Crops Res 60:57–80

Greenwood NN, Earnshaw A (1997) Chemistry of the elements, 2nd ed. Butterworth-Heinemann, Oxford. ISBN 0-7506-3365-4

Grill E, Winnacker EL, Zenk MH (1985) Phytochelatins: the principal heavy-metal complexing peptides of higher plants. Science 230:674–676

Grotz N, Fox T, Connolly E (1998) Identification of a family of zinc transporter genes from *Arabidopsis* that respond to zinc deficiency. Proc Natl Acad Sci U S A 95:7220–7224

Grotz N, Guerinot ML (2006) Molecular aspects of Cu, Fe and Zn homeostasis in plants. Biochimica et Biophysica Acta—Mol Cell Res 1763:595–608

Guerinot ML (2000) The ZIP family of metal transporters. Biochim Biophys Acta 1465:190–198

Gustin JL, Loureiro ME, Kim D et al (2009) MTP1-dependent Zn sequestration into shoot vacuoles suggests dual roles in Zn tolerance and accumulation in Zn-hyperaccumulating plants. Plant J 57:1116–1127

Hambidge KM, Krebs NF (2007) Zinc deficiency: a special challenge. J Nutr 137:1101–1105

Hangavel P (2007) Changes in phytochelatins and their biosynthetic intermediates in red spruce (Picearubens Sarg.) cell suspension cultures under cadmium and zinc stress. Plant Cell, Tiss Organ Cult 88:201–216

Haslett BS, Reid RJ, Rengel Z et al (2001) Zinc mobility in wheat: uptake and distribution of zinc applied to leaves or roots. Ann Bot 87(3):379–386

He XH, Nara K (2007) Element biofortification: can mycorrhizas potentially offer a more effective and sustainable pathway to curb human malnutrition? Trends Plant Sci 12(8):331–333

Holleman AF, Wiberg E, Wiberg N (1985) "Zink" (in German). Lehrbuch der Anorganischen Chemie (91–100 ed.). Walter de Gruyter, Berlin, pp 1034–1041. ISBN 3-11-007511-3

Hotz C, Brown KH (2004) Assessment of the risk of zinc deficiency in populations and options for its control. Food Nutr Bull 25:94–204

Hussain D, Haydon MJ, Wang Y et al (2004) P-type ATPase heavy metal transporters with roles in essential zinc homeostasis in *Arabidopsis*. Plant Cell 16:1327–1339

Institute of Medicine, Food and Nutrition Board (2001) Dietary reference intakes for vitamin A, vitamin K, arsenic, boron, chromium, copper, iodine, iron, manganese, molybdenum, nickel, silicon, vanadium, and zinc

Ishimaru Y, Suzuki M, Kobayashi T et al (2005) OsZIP4, a novel zinc-regulated zinc transporter in rice. J Exp Bot 56(422):3207–3214

Johnson AAT, Kyriacou B, Callahan DL et al (2011) Constitutive overexpression of the OsNAS gene family reveals single-gene strategies for effective iron- and zinc-biofortification of rice endosperm. PLoS One 6(9):e24476

Kalpheck S, Schlunz S, Bergnann L (1995) Synthesis of phytochelatins and homo phytochelatins in *Pisumsativum L*. Plant Physiol 107:515–521

Kapulnik Y (1996) Plant growth promotion by rhizosphere bacteria. In: Waisel Y, Eshel A, Kaffkafi U (eds) Plant roots—the hidden half. pp 769–781. Marcel Dekker, New York

Karley AJ et al (2000) Where do all the ions go? The cellular basis of differential ion accumulation in leaf cells. Trends Plant Sci 5:465–470

Kramer U (2010) Metal hyperaccumulation in plants. Ann Rev Plant Biol 61:517–534

Kupper H, Zhao FJ, McGrath SP et al (1999) Cellular compartmentation of zinc in leaves of the hyperaccumulator *Thlaspi caerulescens*. Plant Physiol 119:305–311

Kupper H, Lombi E, Zhao FJ et al (2000) Cellular compartmentation of cadmium and zinc in relation to other elements in the hyperaccumulator *Arabidopsis halleri*. Planta 212:75–84

Lane BG, Kajioka R, Kennedy TD (1987) The wheat germ Ec protein is a zinc-containing metallothionein. Biochem Cell Biol 65(11):1001–1005

Lombi E, Zhao FJ, Dunham SJ et al (2001) Phytoremediation of heavy metal-contaminated soils: natural hyperaccumulation versus chemically enhanced phytoextraction. J Environ Qual 30(6):1919–1926

Macdiarmidc W, Gaither LA, Eide DJ (2000) Zinc transporters that regulate vacuolar zinc storage in *Sacchromyces cerevisiae*. EMBO J 19:2854–2855

Maestri E, Marmiroli M, Visioli G, Marmiroli N (2010) Metal tolerance and hyperaccumulation: costs and trade-offs between traits and environment. Environ Exp Bot 68:1–13

Maret W (2005) Zinc coordination environments in proteins determine zinc functions. J Trace Elem Med Biol 19:7–12

Martinoia E et al (2007) Vacuolar transporters and their essential role in plant metabolism. J Exp Bot 58:83–102

McGrath D, Tunney H (2010) Accumulation of cadmium, fluorine, magnesium, and zinc in soil after application of phosphate fertilizer for 31 years in a grazing trial. J Plant Nutr Soil Sci 173(4):548–553

Mendez MO, Maier RM (2008) Phytostabilization of mine tailings in arid and semiarid environments—an emerging remediation technology. Environ Health Perspect 116:278–283

Moffett BF, Nicholson FA, Uwakwe NC et al (2003) Zinc contamination decreases the bacterial diversity of agricultural soil. FEMS Microb Ecol 43(1):13–19

Moreau S, Thomson RM, Kaiser BN et al (2002) GmZIP1 encodes a symbiosis-specific zinc transporter in soybean. J Biol Chem 277(7):4738–4746

Nathalieal M, Hassinen NH et al (2001) Enhanced copper tolerance in *Silene vulgaris* (Moench) garcke populations from copper mines is associated with increased transcript levels of a b-type metallothionein gene. Plant Physiol 126:1519–1526

Nishizono H, Ichikawa H, Suziki S et al (1987) The role of the root cell wall in the heavy metal tolerance of *Athyriurn yokoseeme*. Plant Soil 101:15–20

Palmgren MG, Clemens S, Williams LE (2008) Zinc biofortification of cereals: problems and solutions. Trends Plant Sci 13:464–473

Peleg Z, Saranga Y, Yazici A et al (2008) Grain zinc, iron and protein concentrations and zinc-efficiency in wild emmer wheat under contrasting irrigation regimes. Plant Soil 306(1–2):57–67

Pfeiffer WH, McClafferty B (2007) HarvestPlus: breeding crops for better nutrition. Crop Sci 47:S88–S105

Pilon-Smits E (2005) Phytoremediation. Ann Rev Plant Biol 56:15–39

Prasad AS (2007) Zinc: mechanisms of host defense. J Nutr 137:1345–1349

Prasad AS (2008) Zinc in human health: effect of zinc on immune Cells. Mol Med 14(5–6):353–357

Prasad R (2010) Zinc biofortification of food grains in relation to food security and alleviation of zinc malnutrition. Curr Sci 98(10):1300–1304

Prasad AS, Miale A, Farid Z et al (1963) Zinc metabolism in patients with the syndrome of iron deficiency anemia, hepatosplenomegaly, dwarfism and hypogonadism. J Lab Clin Med 61:537–549

Ramesh SA, Choimes S, Schachtman DP (2004) Overexpression of an *Arabidopsis* zinc transporter in *Hordeum vulgare* increases short-term zinc uptake after zinc deprivation and seed zinc content. Plant Mol Biol 54:373–385

Rauser WE (1999) Structure and function of metal chelators produced by plants. Cell Biochem Biophys 31:19–48

Rengel Z (1999) Zinc deficiency in wheat genotypes grown in conventional and chelator-buffered nutrient solutions. Plant Sci 143(2):221–230

Robinson BH, Meblanc L, Petit D et al (1998) The potential of *Thlaspi caerulescens* for phytoremediation of contaminated soils. Plant Soil 203:47–56

Roosens NHCJ, Willems G, Saumitou-Laprade P (2008) Using *Arabidopsis* to explore zinc tolerance and hyperaccumulation. Trends Plant Sci 13:208–215

Römkens P, Bouwman L, Japenga J et al (2002) Potentials and drawbacks of chelate-enhanced phytoremediation of soils. Environ Pollut 116(1):109–121

Sachs J (1865) Handbuch der experimental-physiologie der Pflanzen. In: Hofmeister W (ed) Handbuch der physiologischen botanik. Engelmann, Leipzig

Saier MHJ (1999) A functional phylogenetic classification system for transmembrane solute ransporters. Microbiol Mol Biol Rev 64:354–411

Sanger S, Kneer R, Wanne RG et al (1998) Hyperaccumulation, complexation and distribution of nickle in *Sebertia acuminate*. Phytochemistry 47:339–347

Sayre R, Beeching JR, Cahoon EB et al (2011) The bio-cassava plus program: biofortification of cassava for Sub-Saharan Africa. Annu Rev Plant Biol 62:251–272

Schmidke I, Stephan UW (1995) Transport of metal micronutrients in the phloem of castor bean (*Ricinus communis*) seedlings. Physiol Plant 95:147–153

Schaaf G, Ludewig U, Erenoglu BE et al (2004) ZmYS1 functions as a proton-coupled symporter for phytosiderophore- and nicotianamine-chelated metals. J Biol Chem 279:9091–9096

Schaaf G, Schikora A, Häberle J et al (2005) A putative function for the *Arabidopsis* Fe-phytosiderophore transporter homolog AtYSL2 in Fe and Zn homeostasis. Plant Cell Physiol 46:762–774

Sugarman B (1983) Zinc and infection. Review of infectious diseases 5(1):137–147

Sun Q, Wang XR, Ding SM et al (2005) Effects of interactions between cadmium and zinc on phytochelat in and glutathione production in wheat (*Triticumaestivum L.*). Environ Toxicol 20:195–201

Takahashi M, Terada Y, Nakai I et al (2003) Role of nicotianamine in the intracellular delivery of metals and plant reproductive development. Plant Cell 15:1263–1280

Tandy S, Bossart K, Mueller R et al (2004) Extraction of heavy metals from soils using biodegradable chelating agents. Environ Sci Technol 38:937–944

Tassi E, Pouget J, Petruzzelli G et al (2008) The effects of exogenous plant growth regulators in the phytoextraction of heavy metals. Chemosphere 71:66–73

Tolcin AC (2011) "Mineral commodity summaries 2009: Zinc". United States Geological Survey. Retrieved 2011-06-06

Vamerali T, Bandiera M, Mosca G (2010) Field crops for phytoremediation of metal-contaminated land: a review. Environ Chem Lett 8:1–17

Van de Mortel JE, Almar Villanueva L, Schat H et al (2006) Large expression differences in genes for iron and zinc homeostasis, stress response, and lignin biosynthesis distinguish roots of *Arabidopsis thaliana* and the related metal hyperaccumulator *Thlaspi caerulescens*. Plant Physiol 142:1127–1147

Vasconcelos M, Datta K, Oliva N et al (2003) Enhanced iron and zinc accumulation in transgenic rice with the ferritin gene. Plant Sci 164:371–378

Verbruggen N, Hermans C, Schat H (2009) Molecularmechanisms of metal hyperaccumulation in plants. New Phytol 181:759–776

Verret F, Gravot A, Auroy P et al (2004) Overexpression of AtHMA4 enhances root-to-shoot translocation of zinc and cadmium and plant metal tolerance. FEBS Lett 576:306–312

von Wiren N, Klair S, Bansal S et al (1999) Nicotianamine chelates both Fe-III and Fe-II. Implications for metal transport in plants. Plant Physiol 119:1107–1114

Walker CLF, Black RF (2009) Global and regional child mortality and burden of disease attributable to zinc deficiency. Eur J Clin Nutr 63:591–597

White PJ, Broadley MR (2009) Biofortification of crops with seven mineral elements often lacking in human diets—iron, zinc, copper, calcium, magnesium, selenium and iodine. New Phytol 182(1):49–84

White PJ, Broadley MR (2011) Physiological limits to zinc biofortification of edible crops. Front Plant Sci 2:1–11

Whiting SN, DeSouza MP, Terry N (2001) Rhizosphere bacteria mobilize Zn for hyperaccumulation by *Thlaspi Caerulescens*. Environ Sci Technol 35:3144–3150

White PJ, Whiting SN, Baker AJM et al (2002) Does zinc move apoplastically to the xylem in roots of *Thlaspi caerulescens*? New Phytol 153:201–207

White PJ, Bradshaw JE, Dale MFB et al (2009) Relationships between yield and mineral concentrations in potato tubers. HortScience 44(1):6–11

Yang XE, Long XX, Ni WZ et al (2002) *Sedum alfredii H*: a new Zn hyperaccumulating plant first found in China. Chin Sci Bull 47:1634–1637

Yang XE, Feng Y, He ZL et al (2005) Molecular mechanisms of heavy metal hyperaccumulation and phytoremediation. J Trace Elem Med Biol 18:339–353

Yilmaz A, Ekiz H, Torun B et al (1997) Effect of different zinc application methods on grain yield and zinc concentration in wheat cultivars grown on zinc-deficient calcareous soils. J Plant Nutr 20(4–5):461–471

Zhao H, Eide D (1996a) The yeast ZRT 1 gene encode the zinc transporter protein of a high-affinity uptake system induced by zinc limitation. Proc Natl Acad Sci 93:2454–2458

Zhao H, Eide D (1996b) The ZRT2 gene encode a low affinity zinc transporter in *Saccharomyces cerevisiae*. J Biol Chem 271:23203–23210

Zhao FJ, McGrath SP (2009) Biofortification and phytoremediation. Curr Opin Plant Biol 12:373–380

Zhao FJ, Lombi E, Breedon T et al (2000) Zinc hyperaccumulation and celluar distribution in *Arabidopsis halleri*. Plant Cell Environ 23:507–514

# Chapter 4
# Biofortification to Struggle Against Iron Deficiency

**Yang Huang, Linxi Yuan and Xuebin Yin**

**Abstract** Iron is an essential micronutrient for human beings, but it is not readily available. Consequently, iron deficiency is a major threat to the health and development of the human populations in the world with more than 2 billion people suffering from iron-deficiency anemia. To alleviate iron deficiency, dietary modification or diversification, iron supplementation or food fortification, and crops biofortification have been adopted. Crops biofortification, achieved through three approaches: agronomic intervention, plant breeding, and genetic engineering, could provide a sustainable and cost-effective solution for iron deficiency in food. Agronomic intervention is a traditional approach to increase the yield and quality of crops. Some researches indicate that the application of nitrogen fertilizer and intercropping, such as maize/peanut, guava/sorghum or maize and chickpea/wheat, can facilitate iron uptake by crops. Plant breeding could improve the level and bioavailability of minerals in staple crops through their natural genetic variation. The variation in iron concentration of wheat, maize, and rice suggests that selective breeding might increase the iron content of these staple foods. The transgenic approach for iron biofortification focuses on improving iron accumulation and bioavailability, or decreasing anti-nutrient contents in crops. Expressing ferritin, an iron storage protein, has achieved great success in enhancing iron concentration in seeds. Studies have shown that cysteine could enhance iron absorption in human bodies and thus, greater iron availability is expected by

Y. Huang · L. Yuan · X. Yin
Advanced Laboratory for Selenium and Human Health, Suzhou Institute for Advanced Study, University of Science and Technology of China,
Suzhou 215123, Jiangsu, China

X. Yin (✉)
School of Earth and Space Sciences, University of Science and Technology of China
(USTC) Hefei 230026, Anhui, China
e-mail: xbyin@ustc.edu.cn

X. Yin and L. Yuan (eds.), *Phytoremediation and Biofortification*,
SpringerBriefs in Green Chemistry for Sustainability,
DOI: 10.1007/978-94-007-1439-7_4, © The Author(s) 2012

increasing the amount of cysteine residues in crop tissues. Reducing antinutrients such as phytic acid can also increase the bioavailability of iron.

**Keywords** Iron deficiency · Biofortification · Agronomic intervention · Plant breeding · Genetic engineering

## 4.1 Iron for Human Health

Iron (Fe) is an essential micronutrient for humans. As both an electron donor and acceptor, Iron plays a key role in many vital metabolic pathways such as the electron transport chain of respiration (Gómez-Galera et al. 2010). Iron can form the catalytic active center of heme, which is found in the oxygen-binding molecules, hemoglobin and myoglobin, and the catalytic center of cytochromes, which carry out redox reactions. As a result, Iron is required for oxygen transport and energy metabolism in the body. Iron also contributes to the catalytic activity of a range of non-heme enzymes such as ribonuclease reductase (WHO/FAO 1998). If Fe intake is inadequate, the amount of hemoglobin in the red blood cells can fall leading to iron-deficiency anemia, with symptoms of tiredness, weakness, feeling cold, and inability to concentrate. Moreover, iron deficiency during childhood and adolescence impairs physical growth, mental development, and learning capacity. In adults, it decreases the capacity to do physical labor. Severe anemia increases the risk of women dying during childbirth.

## 4.2 Iron Deficiency

Although iron is the fourth most abundant element in the earth's crust, it is poorly bioavailable in soil because it binds rapidly to soil particles and forms insoluble complexes under aerobic conditions at neutral or alkaline pH. Moreover, dietary iron bioavailability is low in populations consuming monotonous plant-based diets with little meat (Zimmermann et al. 2005), since the absorption of iron is inhibited by the presence of phytates and polyphenols (Hurrell 2002). Therefore, in plant-based diets, iron absorption is often less than 10 % (Zimmermann et al. 2005; Hurrell 2002). Consequently, iron deficiency is the most common micronutrient deficiency in the world. Iron deficiency not only affects the health and development of people, but also hampers the social and economic development of countries due to physical decline of adults. For instance, it is estimated that the loss in economic productivity due to iron deficiency is more than 3.6 % of the gross national product in China (Ross et al. 2003).

Iron deficiency is a worldwide problem. It has been estimated that more than 2 billion of the world's population are iron deficient (WHO 2002; ACC/SCN 2000). In the UK, 21 % of female teenagers between 11 and 18 years, and 18 % of

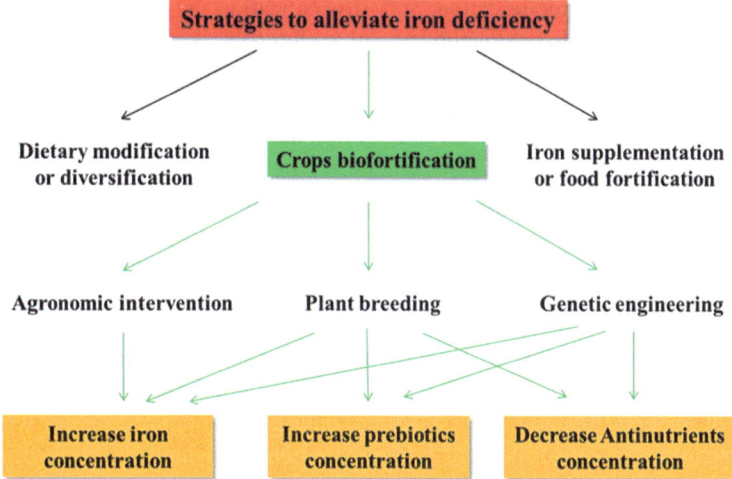

**Fig. 4.1** Schematic flowchart illustrating different strategies that can be used to reduce iron deficiency in humans

women between 16 and 64 years are iron deficient (Heath and Fairweather-Tait 2002). In the USA, 2 % of children between 1 and 2 years (CDC 2002), 9–11 % of nonpregnant women aged between 16 and 49 years are iron deficient, and 2–5 % have iron-deficiency anemia, with more than 2-fold higher frequency in poor, less educated minority populations (Scholl 2005). In pregnant women of low-income families in the USA, the frequency of iron-deficiency anemia in the first, second, and third trimesters is 2, 8, and 27 %, respectively (Scholl 2005). In France, iron deficiency and iron deficiency anemia affect 29 and 4 % of children younger than 2 years (Hercberg et al. 2001). However, iron deficiency is more pronounced in developing countries. WHO estimates that 39 % of children younger than 5 years, 48 % of children between 5 and 14 years, 42 % of all women, and 52 % of pregnant women in developing countries suffer from anemia (WHO/UNICEF/ UNU 2001), with half having iron deficiency anemia (DeMaeyer and Adiels-Tegman 1985). It is estimated that 50 % of pregnant women in developing countries and up to 80 % in South Asia have iron-deficiency anemia.

## 4.3 Strategies to Alleviate Iron Deficiency

Three main strategies can be adopted alone or in combination to alleviate iron deficiency as shown in Fig. 4.1: education combined with dietary modification or diversification, or both, to improve iron intake and bioavailability; iron supple-mentation (provision of iron, usually in higher doses), or iron biofortification on foods and crops (Zimmermann et al. 2007).

Although dietary modification and diversification is the most sustainable approach, change in dietary practices and preferences is generally difficult, and foods that provide highly bioavailable iron (such as meat) are expensive. Iron supplementation can be cost-effective, and can be targeted to high-risk groups (e.g., pregnant women) (Baltussen et al. 2004), but the logistics of distribution and absence of compliance are major limitations. In studies supported by WHO in southeast Asia, iron and folic acid supplementation every week to women of childbearing age improved iron nutrition and reduced iron-deficiency anemia (Cavalli-Sforza et al. 2005). Iron supplementation during pregnancy is advisable in developing countries, where women often enter pregnancy with low iron stores (CDC 2002). However, untargeted iron supplementation in children in tropical countries, mainly in areas of high transmission of malaria, is associated with increased risk of serious infections (Oppenheimer 2001; Gera et al. 2002). In a region of endemic malaria in East Africa, untargeted supplementation with iron (12.5 mg per day) and folic acid in preschool children increased risk of severe illness and even death (Sazawal et al. 2006).

Iron food fortification is probably the most practical, sustainable, and cost-effective long-term solution to mitigate iron deficiency at the national level (Baltdussen et al. 2004; WHO and FAO 2006; Laxminarayan et al. 2006). The overall cost-effectiveness for iron fortification is estimated to be $66–70 per disability-adjusted life year averted (Laxminarayan et al. 2006). However, fortification of foods with iron is more difficult than it is with other nutrients, such as iodine in salt and vitamin A in cooking oil. The most bioavailable iron compounds are soluble in water or weak acid, but often react with other food components to cause off-flavors, and color changes, fat oxidation, or both (Hurrell 2002). Thus, to avoid unwanted sensory changes, less soluble or less absorbed forms of iron are often chosen for fortification.

Biofortification, as a new strategy for modern agricultural production, could be the best choice to solve these problems. The three main biofortification approaches that can be applied, include (i) reducing the concentration in the so-called 'anti-nutrients' [plant metabolites, such as phytic acid (PA) and polyphenols (PP)] that inhibit absorption of dietary iron; (ii) increasing concentrations of other compounds, such as inulin and fructan favoring iron absorption; and (iii) directly increasing the iron concentration (Fig. 4.1).

## 4.4 Molecular Mechanisms of Iron Uptake into Plant Seeds

Plants have evolved two approaches to obtain iron from the soil. In low-iron conditions, non-grasses activate a reduction-based strategy (Strategy I), whereas grasses, such as wheat and corn, adopt a chelation-based strategy (Strategy II) (Fig. 4.2). For rice, both strategies have been adopted which were described by Kim and Guerinot (2007).

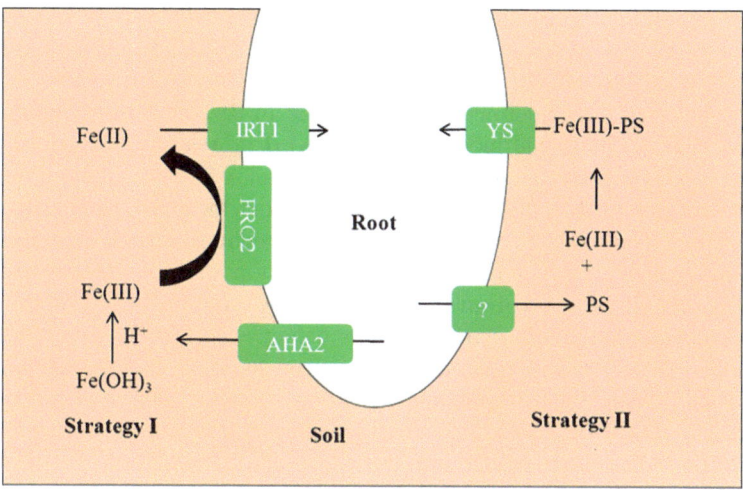

**Fig. 4.2** Two approaches of iron uptake from soil

In Strategy I, non-grasses release protons into the rhizosphere, lowering the pH of the soil solution and increasing the solubility of $Fe^{3+}$, via $H^+$-ATPases of the root plasma membrane (Santi et al. 2005; Santi and Schmidt 2008). After acidification, $Fe^{3+}$ is reduced to $Fe^{2+}$ by a membrane-bound ferric reductase oxidase (FRO), FRO2 (Robinson et al. 1999), one of eight members of the FRO family. Electrons are transferred from $NADH^+$ across four heme groups to iron in the rhizosphere (Robinson et al. 1999). This appears to be the rate-limiting step in iron uptake in *Arabidopsis* (Connolly et al. 2003). Once $Fe^{3+}$ is reduced, $Fe^{2+}$ is transported into the root by iron-regulated transporter 1(IRT1), a member of the zinc-regulated transporter (ZRT)-, IRT-like protein (ZIP) family (Guerinot 2000).

Corn, wheat, and rice use a different mechanism based on chelation, known as Strategy II. In response to iron deficiency, mugineic acid (MA) family PSs are synthesized from L-methionine and released from the root epidermis, perhaps via anionic channels or vesicles (Negishi et al. 2002). PS have high affinity for $Fe^{3+}$ and efficiently bind $Fe^{3+}$ in the rhizosphere. $Fe^{3+}$-PS complexes are then transported into the plant roots via a specific transport system (Curie et al. 2001). It is thought that homologues of the maize (*Zea mays*) yellow stripe 1 (YS1) protein belonging to the oligopeptide transporter (OPT) family are responsible for $Fe^{3+}$-phytosiderophore uptake by Strategy II plants (von Wirén et al. 1995; Curie et al. 2001; Ishimaru et al. 2006; Haydon and Cobbett, 2007; Puig et al. 2007). The YS1 protein is a proton-coupled metal-complex symporter (Schaaf et al. 2004). The chelation strategy is less sensitive to pH than the reduction strategy. Therefore, it is more efficient and allows grass plants to survive under more drastic iron deficiency conditions (Mori et al. 1999).

In addition to having the ability to transport Fe-PS complexes, rice is able to transport $Fe^{2+}$ via OsIRT1 (Ishimaru et al. 2006). In paddy fields, rice can compensate for the lack of effective Fe(III) chelate reductases, and the equilibrium of

$Fe^{3+}/Fe^{2+}$ is shifted in the direction of $Fe^{2+}$ due to the deficiency of oxygen. The adoption of an $Fe^{2+}$ acquisition strategy can be especially advantageous for rice, since rice plants are not very efficient at $Fe^{3+}$ uptake via Strategy II.

After entering the epidermis, iron is required to be bound by chelating compounds, such as citrate and nicotianamine (NA) (Curie et al. 2009; Haydon and Cobbett 2007). Depending on the iron-chelate complex formed, different transport systems are involved in distributing iron throughout the plant Fe-chelator complexes which then move through intercellular connections into the stele along the diffusion gradient. Fe(III)-citrate is the major form of iron present in xylem exudates, and citrate is involved in Fe long distance transport from root to shoot (Grotz and Guerinot 2006). Ferric reductase defective 3 (FRD3), a multidrug and toxin efflux (MATE) family member, is localized to be the plasma membrane of cells in the pericycle and vasculature (Green and Rogers 2004) and functions in iron translocation from roots to shoots by loading citrate into the xylem (Durrett et al. 2007). After citrate loaded into the xylem and chelates iron, Fe(III)–citrate complexes are either taken up at different locations via unidentified transporters or the complexes might be reduced by FROs and $Fe^{2+}$ would then be transported into various cells of the plant. NA is a non-proteogenic amino acid that chelates both $Fe^{2+}$ and $Fe^{3+}$ in addition to other divalent nutrients, such as Cu, Zn, Mn, Co, and Ni (Haydon and Cobbett 2007). NA is synthesized and used in all plants, regardless of their iron uptake strategy, and is a precursor of MA, a PS that is only found in graminaceous plants (Curie et al. 2009; Haydon and Cobbett 2007). NA is structurally similar to PSs and chelates iron for intercellular transport in the phloem. YS-like (YSL) family members are thought to transport metal–NA complexes (Curie et al. 2009; Haydon and Cobbett 2007). There are eight YSLs in *Arabidopsis*, and their proposed functions have been recently reviewed (Curie et al. 2009). YSL1 and YSL3 are suggested to be involved in mobilizing metals, including iron, from leaves for use in developing seeds (Waters et al. 2006).

Presumably, members of the ZIP family are responsible for $Fe^{2+}$ uptake by shoot cells. Members of the natural resistance-associated macrophage protein (NRAMP) family are not thought to be responsible for iron uptake from the soil, but have been implicated in iron homeostasis within plant cells. In particular, NRAMP3 and NRAMP4 are thought to facilitate $Fe^{2+}$ release from the vacuole (Thomine et al. 2003; Grosset al. 2003; Hall and Williams 2003; Lanquar et al. 2005; Grotz and Guerinot 2006; Puig et al. 2007), opposing the activity of the vacuolar iron transporter 1 (VIT1) protein which catalyses iron influx to the vacuole (Kim et al. 2006). In leaves of plants overloaded with iron, and in some seeds, iron can accumulate as Fe-chelates in the vacuole (Pich et al. 2001; Lanquar et al. 2005; Kim et al. 2006). However, under most environmental conditions, the majority of cellular iron is located in the plastid, where it is associated with the Fe-storage protein ferritin (Briat et al. 1999; Petit et al. 2001). The permease in chloroplasts 1 (PIC1) protein is thought to transport iron from the cytoplasm into the plastid (Duy et al. 2007). It is thought that yellow stripe-like (YSL) proteins, and related OPTs, load and unload $Fe^{2+}$-nicotianamine ($Fe^{2+}$-NA) complexes into and out of the phloem for iron redistribution within the plant.

**Table 4.1** Iron-biofortified crops

| Crops | Iron[a] (mg kg$^{-1}$, DW) | References |
|---|---|---|
| *Agronomic intervention* | | |
| Peanut | 363 | Zuo et al. (2000) |
| Wheat | 39 | Aciksoz et al. (2011) |
| Maize | 1630 | Rahman et al. (2011) |
| *Plant breeding* | | |
| Common bean | 85 | Blair et al. (2009) |
| Wheat | 63 | Neelam et al. (2011) |
| Maize | 715 | Pixley et al. (2011) |
| Pearl millet | 5666 | Velu et al. (2011) |
| *Genetic engineering* | | |
| Lettuce | 398 | Goto et al. (2000) |
| Rice | 19 | Nandi et al. (2002) |
| Maize | 35 | Drakakaki et al. (2005) |
| Rice | 16 | Qu et al. (2005) |

[a] Iron concentrations in edible parts of crops

## 4.5 Iron-Biofortified Crops

Biofortification focuses on enhancing the mineral nutritional qualities of crops at source by increasing both mineral levels and their bioavailability in the edible part of staple crops. This can be achieved through agronomic intervention, plant breeding, or genetic engineering (Table 4.1), whereas only plant breeding and genetic engineering can influence mineral bioavailability (Gómez-Galera et al. 2010).

### 4.5.1 Agronomic Intervention

In the soil, iron has a very low mobility, since it binds rapidly to soil particles, when applied as fertilizer in the form of $FeSO_4$, resulting in the conversion of Fe(II) into Fe(III), rendering it unavailable for plant absorption (Frossard et al. 2000). For this reason, prebiotic, such as chelates and nitrogen, are often used along with soil iron fertilizers (Shuman 1998; Rengel et al. 1999). In poor soils lacking the macronutrients (e.g., nitrogen, phosphorus, and potassium), the application of NPK fertilizers can promote the capture of iron, although this also depends on other soil factors like soil pH. More molecular evidences show that remobilization of nitrogen and iron from vegetative tissue to seed is maintained by the similar genetic mechanisms (Uauy et al. 2006; Waters et al. 2009), showing a positive correlation between iron and nitrogen concentrations in grain (Cakmak et al. 2004; Distelfeld et al. 2007). Improving the nitrogen status of plants from low to sufficient high resulted in a 3-fold increase in shoot iron accumulation

(Aciksoz et al. 2011). In addition, the availability of iron in the rhizosphere can be increased by soil acidification with elemental sulfur (Shuman 1998). This has the added benefit of crop sulfur fertilization. Foliar sprays of $FeSO_4$ or chelates allow the direct uptake of iron through leaves. Recently, several research labs have reported that interspecific root interactions and rhizosphere effects could be linked to improved iron and zinc nutrient uptake in dicot plants by intercropping with graminaceous species in pairings which included maize/peanut, guava/sorghum or maize, and chickpea/wheat. For instance, maize and peanut intercropping was shown to improve iron and zinc accumulation in peanut (Kamal et al. 2000; Zuo et al. 2000; Gunes et al. 2007; Inal et al. 2007).

## 4.5.2 Plant Breeding

Plant breeding programs focus on improving the level and bioavailability of minerals in staple crops using their natural genetic variation (Welch and Graham 2005). Breeding for enhanced concentrations of iron can be divided into the following steps: identification of genetic variability within the range that can influence human nutrition; introgressing this variation into high yielding, stress tolerant genotypes possessing acceptable end-use quality attributes; testing the stability of iron accumulation across the target environment; and large-scale deployment of seed of improved cultivars to farmers.

The variation in iron concentration of wheat, bean, cassava, maize, rice, and yam (Welch et al. 2000; Frossard et al. 2000; Haas et al. 2005; Genc et al. 2005; Nestel et al. 2006) suggests that selective breeding might increase the iron content of staple foods. The amount of iron in edible tissues in rice varies between 6 and 22 mg $kg^{-1}$, in maize it varies between 10 and 160 mg $kg^{-1}$, and in wheat the range is 15–360 mg $kg^{-1}$ with the higher levels observed in hydroponic cultures (White and Broadley 2005). Such variability can be exploited through breeding programs to produce iron-enriched crop varieties (Tiwari et al. 2009).

Breeding of nutrient-rich staple food crops is indeed the main goal of different international consortia, such as HarvestPlus (http://www.harvestplus.org/) who aim to reduce micronutrient malnutrition (including provitamin A, zinc, and iron) through different biofortification processes (Hotz and McClafferty 2007), including the dissemination of iron-rich bean varieties for Rwanda and Congo and iron-rich pearl millet varieties for India. AgroSalud (www.agrosalud.org) is another international consortium supporting the production and dissemination of iron- and zinc-rich bean and rice varieties in Latin America and the Caribbean. The efficacy of the breeding strategy in isolating bean varieties that provide more available iron, has been already demonstrated in piglets (Tako et al. 2009), although large-scale trials of efficacy in human population still need to be performed.

However, although differences in iron content exist in wheat and rice, most of the iron is removed during the milling process. Thus, iron concentrations in milled wheat can be increased up to 40 mg $kg^{-1}$, the fortification level commonly used in wheat flour (Flour Fortification Initiative 2004). This problem was evident when the effectiveness of a rice cultivar high in iron was tested in a feeding trial in Filipino women consuming either the high-iron rice ($3.21$ mg kg$^{-1}$) or a local variety ($0.57$ mg kg$^{-1}$) for 9 months (Haas et al. 2005). Possibly because the high-iron rice added only an extra 1.5 mg of iron a day to the diet, no clear benefit of iron status was seen. Iron absorption from other cereals and legumes (many of them have high native iron contents) is low because of their high contents of phytate and polyphenols (Donangelo et al. 2003). Donangelo and colleagues (Hurrell et al. 1999) compared iron bioavailability from two varieties of red beans: an iron-rich genotype (containing 65 % extra iron) and a low-density genotype. Only a small amount of iron was absorbed from both cultivars, probably because of their high phytate and polyphenol content. Decrease of the content of these inhibitors in high-iron cultivars might be needed to have positive effects on human nutrition. Genotypes of maize, barley, and rice have been identified that are low-phytic-acid mutants, with phytic acid, the phosphorus content was decreased by two-thirds compared with wild-type (Raboy 2000). Although such reductions might improve iron absorption from diets containing small amounts of meat and ascorbic acid (Tuntawiroon et al. 1990), the phytic acid content might be needed to be lowered by >90 % to increase iron absorption from the monotonous cereal-based diets popular in many developing countries (Hallberg et al. 1989).

## 4.5.3 Genetic Engineering

The transgenic approaches for iron biofortification in the literature have primarily focused on improving iron accumulation and bioavailability, or decreasing anti-nutrient contents in crops. To enhance iron uptake from soil, plants have been developed with a reduction-based strategy and a chelation-based strategy. Therefore, plants have been transformed with genes encoding transporters, reductases, and other enzymes involved in phytosiderophore biosynthesis (Bauer and Bereczky 2003; Ghandilyan et al. 2006). Transgenic tobacco with an enhanced ferric reductase activity has been developed by introducing a reconstructed yeast FREa gene (Oki et al. 1999). The constitutive ferric reduction capacity was also observed when tobacco plants were grown in the presence of sufficient iron, and the iron level in the transgenic plants was 1.7 times higher than that of the wild-type. In an attempt to enhance the ability of rice to cope with iron-deficient conditions, Takahashi et al. (2001) introduced the gene for nicotianamine aminotransferase, one of the genes in the biosynthetic pathway of phytosiderophores into rice and have shown a better resistance to iron chlorosis in the transgenic plants grown in calcareous soil. Furthermore, the average grain yield of the transformants was 4.1-fold higher than that of the control rice plants. Unfortunately, information about mineral content in the seeds was not reported. Seeds from transgenic barley expressing *Arabidopsis* zinc transporter (AtZIP) show higher iron content than the control (Ramesh et al. 2004).

However, it has been shown that the same channels and transporters that bind iron and zinc can also bind undesirable cations, such as cadmium, copper, manganese, or nickel (Clemens 2006), leading to their accumulation in the plant especially if the soil is deficient in Fe (Connolly et al. 2002; Vert et al. 2002). To avoid the potential toxicity, each transformation event must be tested individually to determine how different metals are absorbed. The unintended effects on other metals can provide benefits, e.g., transgenic tobacco plants expressing genes for enhanced Fe accumulation were also more tolerant of high levels of nickel (Douchkov et al. 2005).

Amplification of the natural iron store is a further strategy to increase iron accumulation in seeds. The greatest success has been achieved by expressing ferritin, an iron storage protein that sequesters the mineral and stores it in a bioavailable form. Recombinant soybean ferritin has been expressed in several cereal crops under the control of an endosperm-specific promoter (Goto et al. 1999; Drakakaki et al. 2005; Vasconcelos et al. 2003; Qu et al. 2005) and pea ferritin has been constitutively expressed in rice (Ye et al. 2008). In each case, Fe levels in edible tissues increased, with the highest levels exceeding 35 mg kg$^{-1}$. The higher iron accumulation upon ferritin overexpression could imply that low iron concentrations in the seed may not result from low iron availability for uptake and transport, but rather from a lack of sequestering capacity in the seed. Murray-Kolb et al. (2002) fed rats with diets containing equivalent amounts of iron as either $FeSO_4$ or transgenic ferritin of different bioengineered rice varieties. Rice diets were as effective as the $FeSO_4$ diet in replenishing hematocrit, hemoglobin concentration, and liver iron concentrations. These data suggested that engineering ferritin expression of seeds can contribute to a sustainable solution of global iron deficiency problems. Another breakthrough in iron biofortification is the development of plants expressing lactoferrin, a human iron-binding protein present in milk that also has broad antimicrobial activity (Bethell and Huang 2004). Recombinant human lactoferrin has been expressed in crops, such as potato (Chong and Langridge 2000), and rice (Anzai et al. 2000; Nandi et al. 2002). Nandi et al. (2002) generated transgenic rice plants in which lactoferrin represented up to 0.5 % by weight of the dehusked grain. They suggested that direct fortification of infant formula with this lactoferrin-rich rice would be a convenient and cost-effective strategy to address childhood diarrhea. Alternatively, lactoferrin could be purified from the transgenic tissues (cell cultures or whole plants) and used directly as a supplement in oral rehydration solutions (Bethell and Huang 2004).

The amount of iron absorbed from a diet can also be increased by improving its bioavailability. This can be achieved by reducing antinutrients such as phytic acid. In vitro studies of the addition of *Aspergillus niger phytase* at pH and temperature conditions similar to that of the stomach could completely degrade phytate (Tuerk and Sandberg 1992). A large number of studies have shown that *A. niger phytase* can be synthesized efficiently in transgenic plants, such as canola, tobacco, and soybean (Brinch-Pedersen et al. 2000). However, *A. niger phytase* is inactivated at temperatures higher than 60 °C. Therefore, this phytase is not useful in transgenic cereals for human consumption, because cooking procedures typically inactivate this enzyme. One solution would be the activation of the transgenic phytase before cooking, during seed development, seed storage, or other processes. An alternative that has been explored is the use of a thermo-tolerant phytase from *Aspergillus fumigatus* (Wyss et al. 1998) in transgenic rice seeds (Lucca et al. 2001). *A. fumigatus phytase* displays a high resistance to heat inactivation, similar to enzymes from thermophilic organisms, because of its ability to refold properly after denaturation. The isolated fungal enzyme was boiled together with rice flour and a residual 59 % phytase activity was observed. However, expression of the heat tolerant *A. fumigatus phytase* in rice endosperm resulted in a strong reduction of the enzyme activity after cooking (Lucca et al. 2001).

Studies have shown that cysteine is the only free amino acid that enhances iron absorption in human bodies (Glahn and Van Campen 1997). Therefore, a greater iron availability is expected by increasing the amount of cysteine residues in crop tissues. A group of cysteine-rich, low molecular weight polypeptides are the metallothionein proteins (MTs) (Cobbett and Goldsbrough 2002). Genes encoding MT-like proteins have been identified in different plant species. Lucca and colleagues (Lucca et al. 2001) overexpressed the cysteine-rich MT gene in rice, as a result, the cysteine content of the soluble seed protein increased about 7-fold.

## 4.6 Summary

Eradicating iron deficiency is a major future goal of the whole world. To solve the problem, a feasible and effective strategy requires effective collaboration among scientists of different disciplines (e.g., plant physiology, biotechnology, human nutrition, epidemiology, and medical care). Dietary modification or diversification, iron supplementation, and food fortification are the current measures to combat iron deficiency. However, they all have their own disadvantages and technical or practical barriers. For instance, the ultimate solution is dietary diversification, but this is not immediately practical. It is evident that crop biofortification is the best approach to alleviate iron deficiency. More researches are needed to ensure successful results in fields, such as the molecular mechanisms of iron absorption and metabolism, antinutrients, and prebiotics.

## References

ACC/SCN (2000) Fourth report on the world nutrition situation. ACC/SCN & International Food Policy Research Institute, Geneva

Aciksoz SB, Yazici A, Ozturk L et al (2011) Biofortification of wheat with iron through soil and foliar application of nitrogen and iron fertilizers. Plant Soil 349:215–225

Anzai H, Takaiwa F, Katsumata K (2000) Production of human lactoferrin in transgenic plants. In: Shimazaki K, Tsuda H, Tomita M, Kuwata T, Perraudin J (eds) Lactoferrin: structure, function and application. Elsevier, Amsterdam, pp 265–271

Baltussen R, Knai C, Sharan M (2004) Iron fortification and iron supplementation are cost-effective interventions to reduce iron deficiency in four subregions of the world. J Nutr 134:2678–2684

Bauer P, Bereczky Z (2003) Gene networks involved in iron acquisition strategies in plants. Agronomie 23:447–454

Bethell DR, Huang JM (2004) Recombinant human lactoferrin treatment for global health issues: iron deficiency and acute diarrhea. Biometals 17:337–342

Blair MW, Astudillo C, Grusak MA et al (2009) Inheritance of seed iron and zinc concentrations in common bean (*Phaseolus vulgaris L.*). Mol Breed 23:197–207

Briat JF, Lobréaux S, Grignon N et al (1999) Regulation of plant ferritin synthesis: how and why. Cell Mol Life Sci 56:155–166

Brinch-Pedersen H, Olesen A, Rasmussen S et al (2000) Generation of transgenic wheat (*Triticum aestivum L.*) for constitutive accumulation of an *Aspergillus phytase*. Mol Breed 6:195–206

Cakmak I, Torun A, Millet E et al (2004) Triticum dicoccoides: an important genetic resource for increasing zinc and iron concentration in modern cultivated wheat. Soil Sci Plant Nutr 50:1047–1054

Cavalli-Sforza T, Berger J, Smitasiri S et al (2005) Weekly iron-folic acid supplementation of women of reproductive age: impact overview, lessons learned, expansion plans, and contributions toward achievement of the millennium development goals. Nutr Rev 63:S152–S158

CDC (2002) Iron deficiency-United States, 1999–2000. MMWR Morb Mortal Wkly Rep 51:897–899

Chong DK, Langridge WH (2000) Expression of full-length bioactive antimicrobial human lactoferrin in potato plants. Transgenic Res 9:71–78

Clemens S (2006) Toxic metal accumulation, responses to exposure and mechanisms of tolerance in plants. Biochimie 88:1707–1719

Cobbett C, Goldsbrough P (2002) Phytochelatins and metal-lothioneins: roles in heavy metal detoxification and homeostasis. Annu Rev Plant Biol 53:159–182

Connolly EL, Fett JP, Guerinot ML (2002) Expression of the IRT1 metal transporter is controlled by metals at the levels of transcript and protein accumulation. Plant Cell 14:1347–1357

Connolly EL, Campbell NH, Grotz N et al (2003) Overexpression of the FRO2 ferric chelate reductase confers tolerance to growth on low iron and uncovers posttranscriptional control. Plant Physiol 133:1102

Curie C, Panaviene Z, Loulergue C et al (2001) Maize yellow stripe1 encodes a membrane protein directly involved in Fe(III) uptake. Nature 409:346–349

Curie C, Cassin G, Couch D et al (2009) Metal movement within the plant: contribution of nictotianamine and yellow stripe 1-like transporters. Ann Bot (London) 103:1–11

DeMaeyer E, Adiels-Tegman M (1985) The prevalence of anaemia in the world. World Health Stat Q 38:303–316

Distelfeld A, Cakmak I, Peleg Z et al (2007) Multiple QTL-effects of wheat Gpc-B1 locus on grain protein and micronutrient concentrations. Physiol Plant 129:635–643

Donangelo CM, Woodhouse LR, King SM et al (2003) Iron and zinc absorption from two bean (*Phaseolus vulgaris L.*) genotypes in young women. J Agric Food Chem 51:37–43

Douchkov D, Gryczka C, Stephan UW et al (2005) Ectopic expression of nicotianamine synthase genes results in improved iron accumulation and increased nickel tolerance in transgenic tobacco. Plant Cell Environ 28:365–374

Drakakaki G, Marcel S, Glahn RP et al (2005) Endosperm-specific co-expression of recombinant soybean ferritin and Aspergillus phytase in maize results in significant increases in the levels of bioavailable iron. Plant Mol Biol 59:869–880

Durrett TP, Gassmann W, Rogers EE (2007) The FRD3-mediated efflux of citrate into the root vasculature is necessary for efficient iron translocation. Plant Physiol 144:197–205

Duy D, Wanner G, Meda AR et al (2007) PIC1, an ancient permease in *arabidopsis* chloroplasts, mediates iron transport. Plant Cell 19:986–1006

Flour Fortification Initiative (2004) Wheat flour Fortification: current knowledge and practical applications. Summary report of an international technical workshop. Cuernavaca, December 1–3. http://www.sph.emory.edu/wheatflour. Accessed 15 June 2007

Frossard E, Bucher M, Machler F et al (2000) Potential for increasing the content and bioavailability of Fe, Zn and Ca in plants for human nutrition. J Sci Food Agr 80:861–879

Genc Y, Humphries JM, Lyons GH et al (2005) Exploiting genotypic variation in plant nutrient accumulation to alleviate micronutrient deficiency in populations. J Trace Elem Med Biol 18:319–324

Gera T, Sachdev HP (2002) Effect of iron supplementation on incidence of infectious illness in children: systematic review. BMJ 325:1142

Ghandilyan A, Vreugdenhil D, Aarts MGM (2006) Progress in the genetic understanding of plant iron and zinc nutrition. Physiol Plant 126:407–417

Glahn RP, Van Campen DR (1997) Iron uptake is enhanced in Caco-2 cell monolayers by cysteine and reduced cysteinyl glycine. J Nutr 127:642–647

Gómez-Galera S, Rojas E, Sudhakar D et al (2010) Critical evaluation of strategies for mineral fortification of staple food crops. Transgenic Res 19:165–180

Goto F, Yoshihara T, Shigemoto N et al (1999) Iron fortification of rice seeds by the soybean ferritin gene. Nat Biotechnol 17:282–286

Goto F, Yoshihara T, Saiki H (2000) Iron accumulation and enhanced growth in transgenic lettuce plants expressing the iron-binding protein ferritin. Theor Appl Genet 100:658–664

Green LS, Rogers EE (2004) FRD3 controls iron localization in *Arabidopsis thaliana*. Plant Physiol 136:2523–2531

Gross J, Stein RJ, Fett-Neto AG et al (2003) Iron homeostasis related genes in rice. Genet Mol Biol 26:477–497

Grotz N, Guerinot ML (2006) Molecular aspects of Cu, Fe and Zn homeostasis in plants. Biochim Biophys Acta 1763:595–608

Guerinot ML (2000) The ZIP family of metal transporters. Biochim Biophys Acta 1465:190–198

Gunes A, Inal A, Adak MS et al (2007) Mineral nutrition of wheat, chickpea and lentil as affected by intercropped cropping and soil moisture. Nutr Cycl Agroecosyst 78:83–96

Haas JD, Beard JL, Murray-Kolb LE et al (2005) Iron-biofortified rice improves the iron stores of nonanemic Filipino women. J Nutr 135:2823–2830

Hall JL, Williams LE (2003) Transition metal transporters in plants. J Exp Bot 54:2601–2613

Hallberg L, Brune M, Rossander L (1989) Iron absorption in man: ascorbic acid and dose-dependent inhibition by phytate. Am J Clin Nutr 49:140–144

Haydon MJ, Cobbett CS (2007) Transporters of ligands for essential metal ions in plants. New Phytol 174:499–506

Heath AL, Fairweather-Tait SJ (2002) Clinical implications of changes in the modern diet: iron intake, absorption and status. Best Pract Res Clin Haematol 15:225–241

Hercberg S, Preziosi P, Galan P (2001) Iron deficiency in Europe. Public Health Nutr 4:537–545

Hotz C, McClafferty B (2007) From harvest to health: challenges for developing biofortified staple foods and determining their impact on micronutrient status. Food Nutr Bull 28:S271–S279

Hurrell RF (2002) How to ensure adequate iron absorption from iron-fortified food. Nutr Rev 60:S7–S15

Hurrell RF, Reddy M, Cook JD (1999) Inhibition of non-haem iron absorption in man by polyphenolic-containing beverages. Br J Nutr 81:289–295

Inal A, Gunes A, Zhang FS et al (2007) Peanut/maize intercropping induced changes in rhizosphere and nutrient concentrations in shoots. Plant Physiol Biochem 45:350–356

Ishimaru Y, Suzuki M, Tsukamoto T et al (2006) Rice plants take up iron as an $Fe^{3+}$-phytosiderophore and as $Fe^{2+}$. Plant J 45:335–346

Kamal K, Hagagg L, Awad F (2000) Improved Fe and Zn acquisition by guava seedlings grown in calcareous soils intercropped with graminaceous species. J Plant Nutr 23:2071–2080

Kim SA, Guerinot ML (2007) Mining iron: iron uptake and transport in plants. FEBS Lett 581:2273–2280

Kim SA, Punshon T, Lanzirotti A et al (2006) Localization of iron in *Arabidopsis* seed requires the vacuolar membrane transporter VIT1. Science 314:1295–1298

Lanquar V, Lelièvre F, Bolte S et al (2005) Mobilization of vacuolar iron by AtNRAMP3 and AtNRAMP4 is essential for seed germination on low iron. EMBO J 24:4041–4051

Laxminarayan R, Mills AJ, Breman JG et al (2006) Advancement of global health: key messages from the Disease Control Priorities Project. Lancet 367:1193–1208

Lucca P, Hurrell R, Potrykus I (2001) Genetic engineering approaches to improve the bioavailability and the level of iron in rice grains. Theor Appl Genet 102:392–397

Mori S (1999) Iron acquisition by plants. Curr Opin Plant Biol 2:250–253

Murray-Kolb LE, Takaiwa F, Goto F et al (2002) Transgenic rice is a source of iron for iron-depleted rats. J Nutr 132:957–960

Nandi S, Suzuki YA, Huang J et al (2002) Expression of human lactoferrin in transgenic rice grains for the application in infant formula. Plant Sci 163:713–722

Neelam K, Rawat N, Tiwari VK et al (2011) Introgression of group 4 and 7 chromosomes of Ae. peregrina in wheat enhances grain iron and zinc density. Mol Breed 28:623–634

Negishi T, Nakanishi H, Yazaki J et al (2002) cDNA microarray analysis of gene expression during Fe-deficiency stress in barley suggests that polar transport of vesicles is implicated in phytosiderophore secretion in Fe-deficient barley roots. Plant J 30:83–94

Nestel P, Bouis HE, Meenakshi JV et al (2006) Biofortification of staple food crops. J Nutr 136:1064–1067

Oki H, Yamaguchi H, Nakanishi H et al (1999) Induction of the reconstructed yeast ferric reductase gene, refre1, into tobacco. Plant Soil 215:211–220

Oppenheimer SJ (2001) Iron and its relation to immunity and infectious disease. J Nutr 131:616S–633S

Petit J-M, Briat J-F, Lobréaux S (2001) Structure and differential expression of the four members of the *Arabidopsis thaliana* ferritin gene family. Biochem J 359:575–582

Pich A, Manteuffel R, Hillmer S et al (2001) Fe homeostasis in plant cells: does nicotianamine play multiple roles in the regulation of cytoplasmic Fe concentration? Planta 213:967–976

Pixley KV, Palacios-Rojas N, Glahn RP (2011) The usefulness of iron bioavailability as a target trait for breeding maize (*Zea mays L.*) with enhanced nutritional value. Field Crop Res 123:153–160

Puig S, Andrés-Colás N, García-Molina A et al (2007) Copper and iron homeostasis in *Arabidopsis*: responses to metal deficiencies, interactions and biotechnological applications. Plant Cell Environ 30:271–290

Qu LQ, Yoshihara T, Ooyama A et al (2005) Iron accumulation does not parallel the high expression level of ferritin in transgenic rice seeds. Planta 222:225–233

Raboy V (2000) Low-phytic-acid grains. Food Nutr Bull 21:423–427

Rahman MM, Soaud AA, AL Darwish FH et al (2011) Growth and nutrient uptake of maize plants as affected by elemental sulfur and nitrogen fertilizer in sandy calcareous soil. Afr J Biotechnol 10:12882–12889

Ramesh S, Choimes S, Schachtman D (2004) Over-expression of an *Arabidopsis* zinc transporter in *Hordeum vulgare* increases short term zinc uptake after zinc deprivation and seed zinc content. Plant Mol Biol 54:373–385

Rengel Z, Batten GD, Crowley DE (1999) Agronomic approaches for improving the micronutrient density in edible portions of field crops. Field Crop Res 60:27–40

Robinson NJ et al (1999) A ferric-chelate reductase for iron uptake from soils. Nature 397:694–697

Ross J, Chen C, He W et al (2003) Effects of malnutrition on economic productivity in China as estimated by profiles. Biomed Environ Sci 16:195–205

Santi S, Schmidt W (2008) Laser microdissection-assisted analysis of the functional fate of iron deficiency-induced root hairs in cucumber. J Exp Bot 59:697–704

Santi S, Cesco S, Varanini Z et al (2005) Two plasma membrane $H^+$-ATPase genes are differentially expressed in iron-deficient cucumber plants. Plant Physiol Biochem 43:287–292

Sazawal S, Black RE, Ramsan M et al (2006) Effects of routine prophylactic supplementation with iron and folic acid on admission to hospital and mortality in preschool children in a high malaria transmission setting: community-based, randomised, placebo-controlled trial. Lancet 367:133–143

Schaaf G, Ludewig U, Erenoglu BE et al (2004) ZmYS1 functions as a proton-coupled symporter for phytosiderophore- and nicotianamine-chelated metals. J Biol Chem 279:9091–9096

Scholl TO (2005) Iron status during pregnancy: setting the stage for mother and infant. Am J Clin Nutr 81:1218–1222

Shuman LM (1998) Micronutrient fertilizers. J Crop Prod 1:165–195

Takahashi M, Nakanishi H, Kawasaki S et al (2001) Enhanced tolerance of rice to low iron availability in alkaline soils using barley nicotianamine amino-transferase genes. Nat Biotechnol 19:466–469

Tako E, Lapparra JM, Glahn RP et al (2009) Biofortified black beans in a maize and bean diet provide more available iron to piglets than standard black beans. J Nutr 139:305–309

Thomine S, Lelièvre F, Debarbieux E et al (2003) AtNRAMP3, a multispecific vacuolar metal transporter involved in plant responses to iron deficiency. Plant J 34:685–695

Tiwari VK, Rawat N, Chhuneja P et al (2009) Mapping of quantitative trait loci for grain iron and zinc concentration in diploid A genome wheat. J Hered 100:771–776

Tuerk S, Sandberg A-S (1992) Phytate hydrolysis during bread-making: effect of phytase addition. J Cereal Sci 15:281–294

Tuntawiroon M, Sritongkul N, Rossander-Hulten L et al (1990) Rice and iron absorption in man. Eur J Clin Nutr 44:489–497

Uauy C, Distelfeld A, Fahima T et al (2006) ANAC gene regulating senescence improves grain protein, zinc, and iron content in wheat. Science 314:1298–1301

Vasconcelos M, Datta K, Oliva N et al (2003) Enhanced iron and zinc accumulation in transgenic rice with the ferritin gene. Plant Sci 164:371–378

Velu G, Rai KN, Muralidharan V et al (2011) Gene effects and heterosis for grain iron and zinc density in pearl millet (Pennisetum glaucum (L.) R. Br). Euphytica 180:251–259

Vert G, Grotz N, Dédaldéchamp F et al (2002) IRT1, an Arabidopsis transporter essential for iron uptake from the soil and for plant growth. Plant Cell 14:1223–1233

von Wirén N, Römheld V, Shioiri T et al (1995) Competition between microorganisms and roots of barley and sorghum for iron accumulated in the root apoplasm. New Phytol 130:511–521

Waters BM, Chu HH, DiDonato RJ et al (2006) Mutations in Arabidopsis Yellow Stripe-Like1 and Yellow Stripe-Like3 reveal their roles in metal ion homeostasis and loading of metal ions in seeds. Plant Physiol 141:1446–1458

Waters BM, Uauy C, Dubcovsky J et al (2009) Wheat (Triticum aestivum) proteins regulate the translocation of iron, zinc, and nitrogen compounds from vegetative tissues to grain. J Exp Bot 60:4263–4274

Welch RM, Graham RD (2005) Agriculture: the real nexus for enhancing bioavailable micronutrients in food crops. J Trace Elem Med Biol 18:299–307

Welch RM, House WA, Beebe S et al (2000) Genetic selection for enhanced bioavailable levels of iron in bean (Phaseolus vulgaris L.) seeds. J Agric Food Chem 48:3576–3580

White PJ, Broadley MR (2005) Biofortifying crops with essential mineral elements. Trends Plant Sci 10:586–593

WHO (2002) World Health Report 2002. Reducing Rishs, Promoting Healthy Life. WHO, Geneva

WHO and FAO (2006) Guidelines on food fortification with micronutrients. In: Allen L, de Benoist B, Dary O, Hurrell R (eds). WHO, Geneva

WHO/FAO (1998) Vitamin and mineral requirements in human nutrition: report of a joint FAO/WHO expert consultation, 2nd edn. Bangkok, Thailand. 21–30 Sept 1998

WHO/UNICEF/UNU (2001) Iron deficiency anemia assessment, prevention, and control. WHO, Geneva

Wyss M, Pasamontes L, Remy R et al (1998) Comparison of the thermostability properties of three acid phosphatases from molds: *A. fumigatus phytase*, *A. niger phytase*, and *A. niger* PH 2.5 acid phosphatase. Appl Environ Microbiol 64:4446–4451

Ye HX, Li M, Guo ZJ et al (2008) Evaluation and application of two high-iron transgenic rice lines expressing a pea ferritin gene. Rice Sci 15:51–56

Zimmermann MB, Hurrell RF (2007) Nutritional iron deficiency. Lancet 370:511–520

Zimmermann MB, Chaouki N, Hurrell RF (2005) Iron deficiency due to consumption of a habitual diet low in bioavailable iron: a longitudinal cohort study in Moroccan children. Am J Clin Nutr 81:115–121

Zuo Y, Zhang F (2009) Iron and zinc biofortification strategies in dicot plants by intercropping with gramineous species. A review. Agron Sustain Dev 29:63–71

Zuo YM, Zhang FS, Li XL et al (2000) Studies on the improvement in iron nutrition of peanut by intercropping maize on a calcareous soil. Plant Soil 220:13–25

# Chapter 5
# Phytoremediation of Cadmium and Copper: Contaminated Soils

**Yuanyuan Zhu, De Bi, Linxi Yuan and Xuebin Yin**

**Abstract** With the development of modern industry and agriculture, the Cadium (Cd) and Copper (Cu) contents in soil have significantly increased. Pollution of Cd and Cu is more serious in a soil–plant environment, so that the remediation of the contaminated environment has been paid more attention. Phytoremediation is to use living green plants to reduce, remove, degrade, or immobilize toxins from contaminated soil, which is an emerging cost-competitive environmental-friendly technology. Recent researches on Cd and Cu hyperaccumulators will be reviewed in this chapter. Future research on the screening Cd and Cu hyperaccumulators, and their molecular mechanisms are necessary for developing phytoremediation.

**Keywords** Cadmium · Copper · Phytoremediation · Hyperaccumulator

## 5.1 Introduction

Phytoremediation is an emerging cost-competitive environmental-friendly technology that cleans up polluted environments through the use of living green plants to reduce, remove, degrade, or immobilize toxins from contaminated soil, water, sediments, or air (Salt et al. 1995). At present, efforts have been focused on using

Y. Zhu · D. Bi · L. Yuan · X. Yin (✉)
Advanced Laboratory for Selenium and Human Health, Suzhou Institute for Advanced
Study, University of Science and Technology of China,
Suzhou 215123, Jiangsu, China
e-mail: xbyin@ustc.edu.cn

X. Yin
School of Earth and Space Sciences, University of Science and Technology of China
(USTC), Hefei 230026, Anhui, China

X. Yin and L. Yuan (eds.), *Phytoremediation and Biofortification*,
SpringerBriefs in Green Chemistry for Sustainability,
DOI: 10.1007/978-94-007-1439-7_5, © The Author(s) 2012

plants and rhizosphere microorganisms to degrade organic pollutants and/or remove toxic heavy metals from contaminated sites. Phytoremediation of contaminated sites has strong public appeal being cost-competitive and environmentally sustainable, compared with other traditional remediation technologies involving excavation or chemical stabilization/conversion in situ. Moreover, phytoremediation could be esthetically pleasing to the public (Wendy et al. 2005).

Phytoremediation includes a set of processes, including phytoextraction, phytostabilization, phytovolatilization, phytofiltration, and phytodegradation. The use of plants to remove contaminants from the environment and accumulate them in above-ground plant tissues is known as phytoextraction. Pollutants can be removed from the contaminated site by harvesting pollutant-accumulated plants (Salt et al. 1998). Metals enter into the xylem through two pathways: one is the apoplastic pathway that soluble metals do not enter through the plasma membrane of epidermis and endodermis cell, but through the cribrate cell wall and the space between cells, and then cross the Casparian strip in endodermis cell; the Casparian strip is a waxy coating which is impermeable to solutes. So, soluble metals need to cross the plasma membrane and then enter into the endodermis cell to the xylem. The other is the symplastic pathway in which soluble metals cross the plasma membrane of epidermis cell and then enter into endodermis cell to the xylem. Once into the xylem, the flow of the xylem sap will transport the metal to shoot or leaf tissues. Once in the shoot or leaf tissues, metals can be stored in various cell types, depending on the species and form of the metal, it can be converted into less toxic forms (to the plant) through chemical and biological transformation or complexation. In leaves or shoots, metals can also be stored in several cellular organelles, such as cell wall, cytosol, and vacuole. Mercury, Se, and As can also be volatilized from leaves (Lombi et al. 2001a).

## 5.2 Cadmium Contamination

Cadmium (Cd) is a heavy metal that is of great concern to the environment. With the development of modern industry and agriculture, the Cd content in soil has significantly increased. Pollution of Cd is more serious in a soil–plant environment, so that the safety of agricultural products has become a serious matter of concern (Davis 1984). In regard to its toxic effects on humans, Cd can cause kidney damage, renal dysfunction, pulmonary emphysema, and osteoporosis (Albert and Hsu 2005). Therefore, soil Cd pollution is one of the most important environmental problems worldwide. The main sources of Cd in soil are from industrial waste discharge, Cd-containing wastewater irrigation and pesticides, herbicides and phosphate fertilizers utilization. In China, Cd contaminated land in 11 provinces and annual production of Cd-contaminated rice (i.e., containing over 1.0 mg Cd kg rice) was about $5 \times 10^4$ tons. A survey conducted in Hunan Province in the early 1990s showed that the agricultural yield was significantly reduced by 5–10 % in Cd-contaminated farmland (Wang 1997). In Japan,

**Table 5.1** Hyperaccumulators of Cd and their accumulation

| Scientific Name | Cd content in shoots or leaves (mg/kg dry weight) | Enrichment coefficient (EC) | Transfer Factor (TF) |
|---|---|---|---|
| *Thlaspi caerulescens* | 164–3,000 | >1 | >1 |
| *Brassica juncea* | >177 | >1 | . > 1 |
| *Sedum alfredii* | >100 | >1 | . > 1 |
| *Solanum nigrum* | 104–125 | 2.68 (average) | 1.04 < TF < 1.27 |
| *Viola baoshanensis* | 465–2310 | 2.38 (average) | 1.32 (average) |
| *Phytolacca acinosa Roxb* | 200–482 | 2.02 < EC < 5.52 | 1.67 < TF < 2.25 |

Cd-contaminated farmland was 472,125 ha, about 82 % of the total heavy metal-contaminated agricultural land (Liao 1993). Albert and Hsu (2005) reported that there were more than 8 % of uncontrolled Cd-contamination hazardous waste sites in the United States. The traditional physical and chemical remediation methods were effective in cleaning up Cd-contaminated soils, but the cost is generally high or very expensive. Therefore, cost-effective remediation technologies are desperately needed for Cd-contaminated sites. Since the late 1990s, phytoremediation has become popular and represents a novel, cost-competitive, and environmental-friendly promising remediation technology for metal-contaminated waters and soils.

## 5.3 Phytoremediation of Cd

Plants that can accumulate over 100 μg/g Cd dry weight (DW) in shoots or leaves can be selected as candidate species for Cd phytoremediation. At present, six kinds of Cd hyperaccumulators have been reported. The species include *Thlaspi. caerulescens* (Baker et al. 1994; Knight et al. 1997; Lombi et al. 2001a, b), *Brassica juncea* (Ebbs et al. 1997), *Sedum alfredii* (Yang et al. 2004; Lu et al. 2010), *Solanum nigrum* (Wei et al. 2010), *Viola baoshanensis* (Liu et al. 2003), and *Phytolacca acinosa Roxb* (Nie 2006) (Table 5.1).

Among the species, *T. caerulescens* accumulated the highest concentrations of Cd in leaves (>164 mg/kg DW). However, the remediation efficiency by *T. caerulescens* is limited due to its slow growth and small biomass production. Although concentrations of Cd in shoots of *B. juncea* (about 100 mg/kg DW) are lower than those of *T. caerulescens*, *B. juncea* can remove more Cd from the contaminated soil due to its larger size of biomass production (Pence et al. 2000). Zhuang et al. (2007) selected the plant species with high biomass production to phytoremediate paddy soils contaminated with Pb, Zn, and Cd. The results indicated that *Viola baoshanensis* accumulated 28 mg Cd/kg DW in shoots, with a bioconcentration factor of 4.8. The total phytoextraction was 0.17 kg Cd per ha, and about 1 % Cd could be removed from the soils, compared with another plant species, *Rumex crispus,* which extracted 0.16 kg Cd per ha (Zhuang et al. 2007). Ji et al. (2011) utilized the Cd hyperaccumulator species, *S. nigrum,* to clean up the farmland soil contaminated by 1.91 mg Cd kg$^{-1}$ in the soil. They found that the planting density had significant effects on plant biomass and Cd accumulation, but *S. nigrum* could accumulate a significant amount of Cd from the soils where the Cd concentrations were relatively low (Ji et al. 2011). However, it should be pointed out that there are some restrictions on phytoremediation application; the stature and biomass of accumulator plants were small and their accumulation ability was limited. Moreover, the phytoremediation time by Cd hyperaccumulators would require many years or a long-term time commitment (Lombi et al. 2001b). Transgenic approaches could break through some of these restrictions to make Cd phytoremediation feasible with overexpression of Cd-accumulation-mediated gene. Bennett et al. (2003) studied that genetically modified *B. juncea* could accumulate 1.5 times more Cd and Zn, compared to wild-type *B. juncea* growing on metal-contaminated soil from a USEPA Superfund site. Overexpression of *E. coli* gene gshI (with a chloroplast targeting sequence) in *B. juncea* could increase the $\gamma$-glutamylcysteine synthetase (ECS) activity by five times compared with wild-type and ECS is important in Cd accumulation and tolerance (Tong et al. 2004). Song et al. (2003) reported that expression of the glutathione-Cd transporter YCF1 in Arabidopsis could significantly increase biomass production and therefore, two times more in Cd uptake.

## 5.4  Cu Contamination

Copper (Cu) is an essential element and enzyme co-factor for oxidases (cytochrome C oxidase, superoxide dismutase) and tyrosinases (Uauy et al. 1998). A suitable amount of Cu in human and animal body are essential for normal biological metabolisms, but excessive will be harmful.

There are two kinds of Cu contamination in soils: one is that derived from weathering and mineralizing of Cu mines and the other is anthropogenic, which is derived from mining, industrial waste, urban waste, sludge, and using of Cu pesticides (such as Bordeaux mixture) (Chen 1996). Some of soils near Cu mines had Cu contents as high as 2,000 mg kg$^{-1}$, which was about 10 times higher than that of non-contaminated soils/sediments (20–30 mg/kg) (Salomons and Forstner 1984). With the development of Chinese industry and modernization of agriculture, Cu and other heavy metals pollution in the environment has become more serious (Ni et al. 2003). Once the soil is contaminated with heavy metals, metal pollutants will accumulate and build up in the soil for a long time. Soil metal pollution affects the biological activity of soil microorganisms and crop growth. Toxic metals finally accumulated in the edible parts of crops will become harmful for human health (Nriagu and Pacyna 1988).

## 5.5 Phytoremediation of Cu

Cultivation and selection of Cu-hyperaccumulating plants is an important means for Cu-phytoremediation. Currently, 37 taxa of Cu hyperaccumulators have been reported, (Song et al. 2004), including Labiatae (5 species), Scrophulariaceae (4 species) and Asteraceae (4 species), Gramineae (3 species), Leguminosae, Cyperaceae and Amaranthaceae (2 species), Caryophyllaceae (1 specie), Convolvulaceae (1 specie), and Tiliaceae (1 specie) (Brooks et al. 1992; Tang et al. 1999). In China, Cu-accumulated plants, such as *Elsholtzia splendens*, *Dianthus superbus*, *Eriophrum comosum*, and *Polygonum microcephalum,* were found (Jiang 2003).

Cu was accumulated in the roots after absorption and avoided excess Cu to interfere with plant photosynthesis and other important physiological processes (Meharg 1993; Baker et al. 1983). However, Hogan and Rauser (1981) and McNair (1981) reported that Cu concentrations in aerial parts of Cu-tolerant plant (*Agrostis gigantea*) are higher than the non-Cu tolerant type. In fact, the cell wall and the vacuole are the most important sites for Cu accumulation, primarily in the chemical form of $Cu^{2+}$. $Cu^{2+}$ is generally not toxic to plants. In a thyrium plant, about 70–90 % Cu was retained in cell wall to prevent from entering protoplast (Branquinho et al. 1997). Vacuolar accumulation of Cu was observed in Cu-tolerant plants, especially in Cu hyperaccumulators. Some lower plants, such as algae, can also accumulate Cu in vacuoles. Many substances in plants, such as enzymes, organic acids (metallothionein, plant metal chelate hormone (PC)), and sugar, can complex Cu and significantly affect the normal physiological processes of plants. Meanwhile, Cu could be detoxified by forming Cu-complexes with these compounds (Grill et al. 1985).

## 5.6 Summary

Phytoremediation is a cost-competitive and sustainable biotechnology for the cleanup of Cd- and Cu-contaminated soils. A few plant species have been identified with the potential for phytoremediation of Cd- and Cu-contaminated environments. Genetic engineering techniques have been applied to develop novel transgenic plants with enhanced ability of metal uptake and accumulation. However, management of Cd- and Cu-contaminated plant waste materials may become a challenging issue. Future research needs to focus on the screening Cd and Cu hyperaccumulators and mechanisms necessary for developing phytoremediation.

# References

Albert TY, Hsu Cheng-non (2005) Electrokinetic remediation of cadmium-contaminated clay. J Environ Eng 131:298–304

Baker AJM, Brooks RR, Pease AJ, Malaisse F (1983) Studies on Cu and cobalt tolerance in three closely related taxa within the genus Silene L. (*Caryophyllaceae*) from Zaïre. Plant Soil 73:377–385

Baker AJM, Reeves RD, Hajar ASM (1994) Heavy metal accumulation and tolerance in British population of the metallophyte *Thlaspi caerulesceas* J. C. Presl (*Brassicaceae*). New Phytol 127:61–68

Bennett LE, Burkhead JL, Hale KL et al (2003) Analysis of transgenic Indian mustard plants for phytoremediation of metal-contaminated mine tailings. J Environ Qual 32:432–440

Branquinho C, Brown DH, Catarino F (1997) The cellular location of Cu in lichens and its effects on membrane integrity and chlorophyll fluorescence. Environ Exp Bot 38:165–179

Brooks RR, Baker AJM, Malaisse F (1992) Copper flowers. Res Explor 8:338–351

Chen HM (1996) Heavy metals in a soil plant system. Science Press, Beijing (in Chinese)

Davis RD (1984) Cadmium in sludge used as fertilizer. Experientia 40:117–126

Ebbs SD, Lasat MM, Brady DJ (1997) Phytoextraction of cadmium and zinc from a contaminated soil. J Environ Qual 26:1424–1430

Grill E, Winnacker EL, Zenk MH (1985) Phytochelatins: the principal heavy-metal complexing peptides of higher plants. Science 230:674–676

Hogan GD, Rauser WE (1981) Role of copper binding, absorption and translocation in copper tolerance of *Agrostis gigantea* Roth. J Exp Bot 32:27–36

Ji PH, Sun TH, Song YF, Ackland ML, Liu Y (2011) Strategies for enhancing the phytoremediation of cadmium-contaminated agricultural soils by *Solanum nigrum*. Environ Pollut 159:762–768

Jiang LY (2003) Copper tolerance and uptake of selected Elsholtzia splendens and phytoremediation of the contaminated soil. PhD thesis, Zhe Jiang University, pp 25 (in Chinese)

Knight B, Zhao FJ, McGrath SP, Shen ZG (1997) Zn and Cadmium uptake by the hyperaccumulator *Thlaspi caerulescens* in contaminated soils and its effects on the concentration and chemical speciation of metals in soil solution. Plant Soil 197:71–78

Liao ZJ (1993) Environment chemistry and biological effects of trace elements. Environmental Science Press, Beijing, pp 301–303 (in Chinese)

Liu W, Shu WS, Lan CY (2003) *Viola baoshanensis*-a new Cd-hyperaccumulator. Chin Sci Bull 48(19):2046–2049 (in Chinese)

Lombi E, Zhao FJ, Dunham SJ, McGrath SP (2001a) Phytoremediation of heavy metal-contaminated soils: natural hyperaccumulation versus chemically enhanced phytoextraction. J Environ Qual 30:1919–1926

Lombi E, Zhao FJ, McGrath SP, Young SD, Sacchi GA (2001b) Physiological evidence for a high-affinity cadmium transporter highly expressed in a *Thlaspi caerulescens* ecotype. New Phytol 149:53–60

Lu LL, Tian SK, Zhang M et al (2010) The role of Ca pathway in Cd uptake and translocation by the hyperaccumulator *Sedum alfredii*. J Hazard Mater 183:22–28

McNair MR (1981) The uptake of copper by plants of *Mimulus guttatus* differing in genotype primarily at a single major copper tolerance locus. New Phytol 88:730–732

Meharg AA (1993) The role of the plasmalemma in metal tolerance in angiosperms. Physiol Plant 88:191–198

Ni CY, Chen YX, Luo YM (2003) Recent advances in research on copper pollution and remediation of soil-plant system. J Zhejiang Univ (Agric Life Sci) 29(3):237–243 (in Chinese)

Nie FH (2006) Cd hyper-accumulator *Phytolacca acinosa* Roxb and Cd-accumulative characteristics. Ecol Environ 15(2):303–306 (in Chinese)

Nriagu JO, Pacyna JM (1988) Quantitative assessment of worldwide contamination of air, water and soils by trace metals. Nature 333:134–139

Pence NS, Larsen PB, Ebbs SD et al (2000) The molecular physiology of heavy metal transport in the Zn/Cd hyperaccumulator Thlaspi caerulescens. Proc Natl Acad Sci U S A 97:4956–4960

Salomons W, Forstner U (1984) Metals in the hydrocycle. Springer-Verlag, Berlin, p 349

Salt DE, Blaylock M, Kumar NPBA et al (1995) Phytoremediation: a novel strategy for the removal of toxic metals from the environment using plants. Biotechnology 13:468–474

Salt DE, Smith RD, Raskin I (1998) Phytoremediation. Annu Rev Plant Physiol Plant Mol Biol 49:643–668

Song WY, Sohn EJ, Martinoia E et al (2003) Engineering tolerance and accumulation of lead and cadmium in transgenic plants. Nat Biotechnol 21:914–919

Song J, Zhao FJ, Luo YM, McGrath SP, Zhang H (2004) Copper uptake by *Elsholtzia splendens* and *Silene vulgaris* and assessment of Copper phytoavailability in contaminated soils. Environ Pollut 128:307–315

Tang SR, Wilke BM, Huang CY (1999) The uptake of copper by plants dominantly growing on copper mining spoils along the Yangtze River, the people's Republic of China. Plant Soil 209:225–232

Tong YP, Kneer R, Zhu YG (2004) Vacuolar compartmentalization: a second-generation approach to engineering plants for phytoremediation. Trends Plant Sci 9:7–9

Uauy R, Olivares M, Gonzalez M (1998) Essentiality of copper in humans. Am J Clin Nutr 67:952–959

Wang KR (1997) Cadmium pollution and treatment countermeasure of farmland in China. Agric Environ Prot 16:274–278 (in Chinese)

Wei SH, Li YM, Zhou QX et al (2010) Effect of fertilizer amendments on phytormediation of Cd-contaminated soil by a newly discovered hyperaccumulator discovered hyperaccumulator *Solanum nigrum L.* J Hazard Mater 176:269–273

Wendy AP, Ivan RB, Elizabeth LR et al (2005) Phytoremediation and hyperaccumulator plants. In: Tamás MJ, Martinoia E (eds) Molecular biology of metal homeostasis and detoxification. Springer-Verlag, Berlin

Yang XE, Long XX, Ye HB et al (2004) Cadmium tolerance and hyperaccumulation in a new Zn-hyperaccumulating plant species (*Sedum alfredii* Hance). Plant Soil 259:181–189

Zhuang P, Yang QW, Wang HB, Shu WS (2007) Phytoextraction of heavy metals by eight plant species. Water Air Soil Pollut 184:235–242